My Journey from Mainframe to AI

A Personal and Professional Guide to Understanding the Past, Present, and Future of Technology

by
Lewin Wanzer

Tourque Press

Printed in the United States of America
First Edition

ISBNs:
Paperback: 979-8-9987073-0-8
Hardcover: 979-8-9987073-1-5
eBook (EPUB): 979-8-9987073-2-2
Cover design by Hasina Afzal
For more information, visit: www.tourqueai.com

Dedications

To God—my source, my guide, my strength.
To my father and mother, who first taught me resilience.
To my wife and kids, who are my greatest reasons why.
To my **extended family—my cousins, aunts, uncles, and all who've walked beside me**—your presence has been a constant foundation through every season of growth.
To my church family at KNI—your prayers, support, and love have carried me through silence, struggle, and breakthrough.
To the musicians who reminded me that creativity is a form of architecture.
To the technologists, mentors, and heroes I've had the honor of working alongside—thank you for the sparks, the late-night calls, the challenges, and the breakthroughs.
And to those who tested me, stretched me, and put me in rooms where failure wasn't an option—
You didn't just shape my opinions on technology. You helped shape the foundation of **Tourque** and **Tourbook**.

This book is the product of all of you. Thank you for being part of the story.

Philippians 4:13
"I can do all things through Christ who strengthens me."

Acknowledgments

I want to thank every company that took a chance on whether as an employee, consultant, or contractor. Each opportunity gave me a front-row seat to the evolution of technology and the chance to grow through hands-on experience, late-night problem solving, and real-world innovation.

To all the teams I've worked with, the platforms I've helped build, and the systems I've been trusted to fix—thank you. This journey has been shaped by every challenge, every expectation, and every line of code.

Special thanks to those who encouraged me to keep going when things were unclear, who believed in my ability to bring value, and who trusted me as a technical resource even when the path forward wasn't obvious.
And yes, those who underestimated me, questioned my worth, or saw me only through the lens of a job title—thank you, too. Your doubt became fuel. Your resistance became refinement. You helped shape the technologist and leader I am today.

Every experience—good and bad—was a teacher. And I'm grateful for them all.

Proverbs 16:3
"Commit to the Lord whatever you do, and He will establish your plans."

Table of Contents

Introduction

This isn't just a book about technology. It's a story about transitions, timing, reinvention—and purpose.

I've spent decades navigating the ever-changing terrain of IT, from mainframes to AI. But underneath all the systems, tools, and processes, one constant has remained: **God has guided me every step of the way**. Even when I didn't see the path ahead, He was already preparing the next phase.

I wrote this book not just to document where we've been in the world of technology, but to offer perspective—and hope—for those facing disruption, reinvention, or uncertainty. Whether you're an IT leader, a creator, a builder, or someone trying to figure out what comes next, I want you to see that there is a divine thread in all of it.

We were made to build. To create. To take what's broken and find a better way. I believe that desire comes from the Creator Himself and when aligned with vision, discipline, and humility, it can spark something that changes not just systems, but lives.

This is my story. But maybe it's yours too.
Welcome to *Lessons from the Field, the Stage, and the Data Center*.

About the Author

Lewin Wanzer is a technologist, musician, and lifelong learner who built his career on hard work, humility, and a deep belief in doing things the right way—even when it's not the easy way. A former bassist and national-level athlete, Lewin brings rhythm, discipline, and collaborative spirit to everything he touches—from supporting early mainframes to architecting AI-driven enterprise systems.

Born into a family with primarily a rich mixed family heritage, Lewin leads with servant values: feed your people first, build systems that serve others, and listen more than you speak. He's consulted with over 100 organizations across industries, trained global teams, and written Implementing SharePoint 2019 as a guide for practical, grounded collaboration.

Today, he is the founder of Tourque AI, an AI Operating System designed to help organizations move beyond chaos and toward purpose. His story is not just about systems—it's about how we build, lead, and live with intention.

Chapter 1:

A World of Mainframes

Lessons from the Field, the Stage, and the Data Center

In 1983, the world was on the verge of a technological revolution—but we weren't quite there yet. I had just graduated from high school. There were no personal computers in homes, no mobile phones, and certainly no internet as we know it today. Technology was still exclusive to large corporations, government agencies, and research institutions.

A year or two before that time, my world wasn't about computers, it was about competition. I played all three major sports: football, basketball, and baseball. In 1982, my baseball team ranked second in the nation among Christian schools. In 1983, our basketball team made it to the national tournament in Maranatha, Wisconsin, and came in second as well. Those experiences weren't just athletic achievements, they were lessons in discipline, endurance, and pressure.

And my coaches? They didn't go easy on us. I remember one football practice where we ran the hill 100 times just because we lost a game. My basketball coach made us run 100 suicides. My baseball coach had me running forty 40-yard dashes and ten laps around the field before practice even started.

Looking back, I didn't love the pain—but I learned to endure it. I learned how to push past fatigue, stay focused under pressure, and keep improving even when I failed. Those same lessons would later show up in my career in technology, where a single symbol or misplaced comma could crash a system—and where resilience and repetition made the difference between frustration and success. Even then, I didn't know it, but God was training me—not just physically, but spiritually. Learning endurance under pressure would prepare me for future valleys I couldn't yet see.

"No discipline seems pleasant at the time, but painful. Later on, however, it produces a harvest of righteousness."
– Hebrews 12:11

During my teens I also was into music. I learned from my brothers early on as one sang, the other played guitar and I was around music all the time. I ended up taking one of his 6 string acoustic guitars and I started to play on it. At first it seemed a little clunky, but I ended up learning many bass lines of songs in the late 70's and 80's. I loved all kinds of music and did not just listen to R&B.

One day, I ran into a friend who introduced me to a drummer—and that drummer introduced me to Rush and the legendary Geddy Lee. I had never heard anything like it. His playing was so advanced, so precise, and so musical that it blew my mind. From that moment on, I was hooked. I made it a personal challenge to learn some of Rush's most complex bass lines. It was not easy—but it elevated my skills to a whole new level.

2

Looking back, that moment taught me a lesson I would carry with me well into my career in technology: growth often comes from unexpected connections. Meeting the drummer led to the discovery of Geddy Lee's music. Similarly, new technological advancements are often introduced through conversations, mentorship, or shared links, opening new opportunities.

The real question is: how do we respond when those doors open? Do we back away because the challenge looks too hard—or do we push through and grow?

Another lesson I took from that time: do not stay stuck in one lane. I was playing mostly R&B and pop before that. Rock was not on my radar—until that introduction. But learning a completely different genre gave me new scales, techniques, and styles. It expanded how I approached my instrument. Today, I can play the bass in six diverse ways, not because I had formal training, but because I never backed away from a challenge. I embraced diversity of thought and style.

And that mindset? It is the same one I bring to my work in IT and AI.
Keep evolving. Keep listening. Keep learning—even when it sounds different.

During this time, my band traveled to schools and various locations around our area and even a hospital where mental patients lived. Although we thought it was weird we talked to a doctor and the people enjoyed coming to play for them. They also said the music was stimulating to them and helped them.

As I got older there were more places I played and different situations I faced. One thing that really stood out for me was the fact that I at some point would get calls from different bands to come and sit in. Sit In is when you do not know the set or the music the bands are going to play but you are expected to play anyway.

Most of the time it was just music from the radio or oldies I would have to remember but overall, I did well at improvising and facing the challenge. After thinking about this in my later years I noticed that I can be fearless to take on challenges. This resonates from my early years as an athlete and as a musician. Soon I will be faced with a challenge of learning business and how computers would shape the course of history.

If you walked into an office or data center in 1983, you wouldn't see desks cluttered with laptops or employees typing away on high-resolution monitors. Instead, you'd see rows of dumb terminals connected to a massive mainframe, where every request had to be processed sequentially.

A mainframe is a potent central computer used primarily by large organizations to process vast amounts of data quickly and reliably. Think of it like the brain of an entire company—handling everything from payroll to inventory to scientific research—while hundreds (or even thousands) of people connect to it through simple terminals. Unlike today's personal computers or smartphones, mainframes were massive, expensive machines that filled entire rooms and required specialized operating environments.

These mainframes were the backbone of computing, and businesses depended on them for finance, payroll, inventory management, and research.

Outside of corporate IT, the world was very different:

- There was no PlayStation, no Xbox, and no CD player. People listened to music on vinyl records, cassette tapes, and 8-track players. The Sony Walkman had just started gaining popularity, allowing people to listen to music on the go.
- No MP3s or streaming services—if you wanted to hear a song, you had to buy a physical record or tape or wait for it to come on the radio.
- No email for everyday communication. Instead, businesses used fax machines and physical mail.
- Star Wars: Return of the Jedi was in theaters, Reagon was president, and Michael Jackson's Thriller dominated the airwaves.
- Cable TV was new; most had only a few television channels. If you missed a TV show, there was no DVR or Netflix—you had to wait for a rerun.
- Arcades were the dominant form of video gaming, with games like Pac-Man, Donkey Kong, and Galago filling rooms full of quarter-hungry machines.
- Computers were mainly seen as business machines; if you wanted one at home, you were likely a hobbyist or a programmer.

Pong and Atari: The Start of Gaming and the Need for Balance

I still remember the excitement of getting Pong, which hit the market in 1972, followed by the Atari 2600, released in 1977. Pong's simple two-paddle tennis-style game was mesmerizing at the time—it was one of the first video games we could play at home. When Atari came along, it felt like the future had arrived, with interchangeable game cartridges and colorful graphics that brought arcade fun into our living rooms.

But after the initial hype, those systems started collecting dust. We naturally drifted back outside to ride bikes, play football, and hang out with friends. Unlike today, screen time didn't dominate our childhood. However, many kids today spend hours glued to screens, missing the benefits of outdoor play and face-to-face interaction. We must break this cycle. Exercise and real-world social experiences are essential for physical health and building confidence, empathy, and resilience. The games will always be there, but the memories we make running around outside with friends last far longer than any high score.

At the time, two visionaries were beginning to reshape the future of technology, but they hadn't yet reached their peak:

Where Were Bill Gates & Steve Jobs in 1983?

- Bill Gates was only 28, leading Microsoft, a small software company. Just two years earlier, they had released MS-DOS, an operating system designed for IBM's first personal computer. At this time, Microsoft was nowhere near the giant it is today—it was just beginning to gain traction in the business world.
- Steve Jobs was 28 as well, running Apple, which had just launched the Lisa computer, an early attempt at a graphical user interface (GUI). However, the Lisa was a failure, costing nearly $10,000 and selling poorly. Jobs and Apple were about to launch the Macintosh in 1984, which would change everything.

At this moment in history, personal computing was in its infancy, and most businesses still relied on mainframes.

6

The dominant force in IT was the mainframe, a massive centralized system that powered everything from government agencies to Fortune 500 enterprises. These machines relied on batch processing, meaning jobs were submitted into a queue and processed sequentially. There was no real-time interactivity, no instant feedback—just waiting. You submitted a job, walked away, and hoped it was successful when you checked back later.

My First Experience with Computing – Entering the World of Mainframes

I graduated high school in 1983 and had no idea I was about to enter the computing industry. I went to a trade school to learn programming but did not finish my certification due to baseball and music. At the time, the company I worked for supported an agency that was one of the few organizations pushing the boundaries of computing, and thanks to my brother, who was a supervisor, I landed a job at a computer tape library warehouse. I didn't know then, but that warehouse job – thanks to my brother – was the spark that would ignite my lifelong journey into tech.

At first, it was just a paycheck, but this experience unknowingly shaped my entire career.

Entering the World of Mainframes

I worked in the tape library, where my company supported significant research and stored satellite data on magnetic tapes. Every request for data meant physically retrieving a tape, mounting it on a drive, and waiting for the system to process it. It was all manual labor, but this was state-of-the-art computing. At the time, no one could imagine computers writing essays or generating images- alone running on a device in your pocket.

"Technicians Managing a Mainframe System – 1950s or 1960s"
Two operators work alongside a towering mainframe computer system, surrounded by magnetic tape drives and control panels. This image captures the scale and complexity of enterprise computing in the pre-PC era, where managing data required physical interaction, precision, and constant monitoring.

Massive Tape Storage & Cataloging Data

My job was to manage tape storage, the physical medium that stored all the company's vital research, satellite data, and computing processes. Unlike today's cloud storage, where data is instantly retrievable, back then, if someone needed data, we had to physically locate the tape, mount it, and process the request.

The process was entirely manual:

- Tapes were stored in massive boxes, stacked on pallets, and categorized by storage numbers.
- Each tape was manually logged into a physical record book and entered an early warehouse management system.
- Requests were made via printed forms or phone calls without digital automation.
- If a tape was needed, I had to retrieve it, mount it onto a drive, and ensure the system processed it correctly.

Finding out if a tape was misfiled or lost was a nightmare. Sometimes, we had to physically search for missing tapes, like losing an essential document in a cluttered filing cabinet.

In some cases, tapes were too large to be stored in regular shelving, so we had high-tier storage that required forklifts to access. These were the equivalent of today's high-capacity storage servers, only ten times more inefficient.

"Mainframe Tape Reel – A Data Storage Workhorse"

Large magnetic tape reels like this one were the backbone of data storage in the mainframe era. Stored in bulky containers and stacked on pallets, these reels held critical research and operational data. Moving and managing them often requires forklifts and precise cataloging to keep the data center running smoothly.

Managing this data wasn't just about storing tapes. It required an elaborate tracking and retrieval system. We had large boxes for big reels and smaller boxes for compact tapes, all stacked onto wooden skids. Each box was numbered and logged into a cataloging system so we could track where tapes were stored.

Every tape had a unique identification number assigned to a box, which was logged into a warehouse system. When a programmer or scientist needs data, a request form is submitted manually by paper form, through the carrier, or via phone in an emergency. In emergencies, a supervisor would have to verify the request.

Then, it was our job to locate the right tape and either:

- Send it to our local center building location
- Ship it via courier to another facility

We even had high-tier storage, which required a forklift to retrieve large stacks of tapes. Metal casings protected some of these tapes to prevent damage, but others were stored in plastic containers, making them more vulnerable. Tapes could be misfiled, lost, or even damaged, like a corrupt file on a modern hard drive.

Printing & Paper Waste

One of the mainframe era's most enormous inefficiencies was paper waste. Nearly every report, request, and data entry process required printing because everything was stored on magnetic tapes instead of screens or hard drives.

Almost every business process required physical printouts because everything was stored on tapes or punched cards. We printed logs, job reports, data records, and warehouse storage lists, which were all stored in physical filing cabinets.

One of our largest weekly printouts was the warehouse log, a massive document listing every stored tape, category, and retrieval location. Unlike today's standard 8.5x11 paper, these reports were printed on:

- Wide-format continuous-feed paper
- Perforated, accordion-folded sheets
- Dot-matrix printers that used ink ribbons

Because reports were frequently updated, massive stacks of unused or outdated paper would pile up, filling entire rooms. Today, we take paperless offices and cloud storage for granted, but printing was necessary back then.

We printed a massive weekly warehouse log to categorize tapes, boxes, and storage locations. These reports weren't printed on the standard 8.5x11 paper we use today. Instead, they were produced on wide-format continuous-feed printer paper with perforated edges and holes on the sides for use in dot-matrix printers.

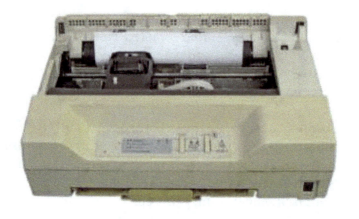

"Dot Matrix Printer with Continuous Feed Paper"
This classic dot matrix printer illustrates the wide-format, perforated paper used for massive print jobs in mainframe environments—often producing reports that stretched for feet at a time. These printers were noisy, slow, and essential to daily operations.

These printouts were thick, heavy stacks of paper that often spanned several feet when unrolled. The process was so inefficient that:

- Stacks of unused paper piled up when reports weren't needed.
- Errors meant reprinting entire sections, wasting even more paper.
- Storage cabinets were filled with paper logs, taking up valuable office space.

Back then, no one thought about saving trees or reducing wastepaper as part of the job. Companies didn't recycle like today, and hundreds of thousands of pages were discarded monthly.

By today's standards, this was an environmental disaster, but it was just how things were done at the time.

How Mainframes Worked

Mainframes were designed for centralized computing, meaning all data was stored and processed in one central location while users accessed it via dumb terminals. These terminals had no processing power; they sent input to the mainframe and displayed the output.

Terminal Connectivity & Speed

All terminals were connected using coaxial cables, which were bulky and slow by today's standards. However, since the displays were text-only with no images, the speed didn't matter, there was no need for high-resolution graphics.

When users logged in, they were greeted with a menu of options. Programmers could submit jobs requiring access to tape-stored data, but if a specific tape weren't available, it had to be retrieved from off-site storage—where I worked.

Memory & Processing Limitations

Unlike modern computers that rely on SSD storage and high-speed RAM, mainframes use magnetic core memory and tape-based storage. Memory was minimal, which meant every program had to be efficiently coded.

For comparison:

- IBM System/360 (1964): Up to 8MB of RAM (costing hundreds of thousands of dollars).
- IBM 3081 (1980s): Up to 32MB of RAM (a breakthrough).
- Today's Smartphones: 8GB+ RAM—250,000 times more than we had back then.

Why Companies Use Mainframes

Despite their massive size and cost, mainframes were highly reliable, secure, and efficient. They were built to last, often running non-stop for decades without failure.

Large corporations and government agencies used mainframes for:

- Banking transactions (processing millions of records daily).
- Airline reservations (e.g., American Airlines' SABRE system).
- Payroll systems for thousands of employees.
- Scientific research and military applications.

The Cost of Mainframes

Owning a mainframe was a significant financial commitment, including the cost of hardware, infrastructure, staffing, and operational expenses.

To put things into perspective:

- IBM 370 mainframes (1970s): $200,000 – $4 million.
- IBM 3081 (1980s): $3.5 million base price, with an additional $500,000 – $1 million per year in maintenance.
- Electricity costs: These machines require massive cooling systems, leading to sky-high electricity bills.

Because of these costs, companies had no choice but to buy IBM products, making IBM the undisputed leader in computing for decades.

Halon Fire Suppression Systems: The High-Cost Guardian of Mainframes

In the 1980s, Halon fire suppression systems were the gold standard for protecting mainframe environments. These systems were designed to quickly extinguish fires without damaging sensitive electronic equipment, a necessity when millions of dollars of computing hardware were at stake. However, while effective, Halon systems were costly to install and maintain.

*Halon gas tanks – The backbone of 1980s
mainframe fire protection*

The Cost and Complexity of Halon Systems

- Installation Costs: Installing a Halon system requires a sealed environment, specialized piping, and dedicated Halon storage tanks, all of which

  add to the costs. A single Halon system could cost hundreds of thousands of dollars, making it one of the most significant expenses in data center construction.

- Maintenance Costs: Halon had to be stored at specific pressures, and leak checks were regularly conducted. The cost of refilling a Halon system after a discharge could be astronomical, especially as Halon became more regulated.
- Testing Hazards: Even testing a Halon system was a logistical nightmare. Unlike modern fire suppression systems allowing localized testing, a Halon test meant deploying gas into the room. This was dangerous—Halon removed oxygen to suppress fires, which meant anyone inside during a discharge risked suffocation.
- Emergency Protocols: Operators were trained to evacuate the room immediately by the sound of an alarm because once Halon was released, you had only seconds to escape before the air became unbreathable.
- Environmental Concerns: Halon was phased out by the 1990s due to its ozone-depleting properties, but it was still the standard in the 1980s. The high cost of replacing it later contributed to companies' transitioning to other fire suppression methods.

Air Conditioning and Humidity Control: The Silent Battle

While fire suppression was a necessary safeguard, temperature and humidity control were daily concerns for data center operators. Mainframes generated immense heat and keeping them within the proper operating temperature was non-negotiable.

Logging Room Temperature and Humidity

- Daily Logs: As operators, we were required to log the temperature and humidity multiple times per shift manually. This was critical because a failure in the cooling system could lead to overheating, system failures, or even permanent hardware damage.
- Temperature Requirements: Mainframe rooms had to be maintained between 60-70°F (15 21 21°C). Any deviation could spell trouble.
- Humidity Control: Humidity levels must stay between 40-55%. If the humidity is too low, static electricity buildup could fry a mainframe board. If it is too high, condensation could form on circuits, leading to short circuits and catastrophic failures.

The Cost of Air Conditioning

- Massive AC Systems: These were not your standard office air conditioners. Large-scale, high-capacity AC units with redundant cooling towers were required.
- Electricity Consumption: Keeping a large data center cool could cost tens of thousands of dollars monthly in electricity alone. In some cases, cooling costs exceeded the cost of running the computers.
- Backup Cooling Systems: Many data centers had multiple cooling units; some even had water-cooled systems to handle excess heat. If an AC unit failed, a backup had to kick in immediately, or systems would overheat in minutes.

The Reality of Working in a Cold Data Center

- Jackets Required: Despite working in a high-tech environment, we often had to wear jackets or sweaters because the rooms were kept at such low temperatures.
- Condensation Risks: Operators had to be careful about bringing in drinks, as the cold air could cause sweating on cups, leading to accidental spills—something you never wanted near a million-dollar system.
- Hearing the Hum: Between the roar of the AC systems and the constant hum of the mainframes, it was never silent. The cooling systems always ran at full blast to keep the environment stable.

Managing a data center in the 1980s wasn't just about operating computers and maintaining an entire ecosystem. Fire suppression, air conditioning, and humidity control were constant concerns.

While effective, the Halon system was looming in every operator's mind due to its dangerous testing procedures and astronomical costs. Meanwhile, temperature and humidity regulation were daily responsibilities, ensuring that mainframes operated at peak efficiency without succumbing to heat or moisture.

Those who worked in these environments remember the cold air, the fear of an unexpected Halon discharge, and the ever-present logs of room conditions. It was a high-stakes world that required precision, vigilance, and a deep respect for the technology we were maintaining.

Those early days working in the tape library warehouse weren't glamorous. Still, they taught me the discipline, structure, and responsibility that would carry me through decades of technological change – from magnetic tapes to machine learning algorithms.

Chapter 2:

Programming in the Mainframe Era – The Building Blocks of Enterprise Computing

In the early 1980s, programming was far from what we know today. There were no drag-and-drop interfaces, cloud-based compilers, or real-time debugging tools. Every line of code had to be meticulously crafted because errors were expensive and time-consuming.

Unlike today's variety of high-level programming languages, coding in the mainframe era was heavily dictated by industry needs and hardware constraints. Developers had to think about memory efficiency, batch processing limits, and storage constraints, as even a single wasted data byte could be costly.

Most programming was text-based and batch-oriented, meaning you wrote the program, submitted it to the system, and then waited—sometimes minutes or hours- to run and produce output. If a mistake was found, the entire process had to be repeated.

The Top Programming Languages of the 1980s

In the early 1980s, there were only a few dominant programming languages, each catering to specific industries and computing needs:

- COBOL (Common Business-Oriented Language)
    - Industry Use: Finance, government, banking, payroll, and large-scale transaction processing.
    - Why It Was Popular: COBOL was designed for business applications, making it the go-to language for financial records, tax processing, and accounting systems.
    - Still Used Today? Yes. Despite being over 60 years old, COBOL still runs much of the world's banking infrastructure, and many government systems still rely on it.
    - Example Use Case: The Social Security Administration and IRS were among the many government agencies that used COBOL to process records and transactions.
- FORTRAN (Formula Translation)
    - Industry Use: Engineering, scientific research, aerospace, NASA, and military applications.
    - Why It Was Popular: FORTRAN was optimized for mathematical and scientific calculations, making it the dominant language for complex computational models and simulations.
    - Still Used Today? Yes, in scientific computing and engineering simulations.
    - Example Use Case: NASA used FORTRAN for spaceflight calculations, satellite modeling, and aerodynamics research.

- Pascal
  - Industry Use: Education, early personal computing, and structured programming applications.
  - Why It Was Popular: Pascal was widely taught in universities and helped students learn structured programming techniques.
  - Still Used Today? Not commonly, but it laid the foundation for modern programming paradigms like Object-Oriented Programming (OOP).
  - Example Use Case: Many early Apple systems and educational environments used Pascal before C and C++ became more widespread.
- C (The Foundation of Modern Programming)
  - Industry Use: Operating systems, embedded systems, networking, and personal computing.
  - Why It Was Popular: C offered low-level hardware control while being more readable than assembly language.
  - Still Used Today? Yes—almost every modern programming language (C++, Java, Python, JavaScript) is based on principles introduced in C.
  - Example Use Case: UNIX operating systems were written in C, making it a core language for system programming.

- RPG (Report Program Generator)
  - Industry Use: Business applications, retail, and manufacturing.
  - Why It Was Popular: RPG was designed for quick report generation and data entry automation on IBM mainframes.
  - Still Used Today? Yes, but mostly in legacy IBM AS/400 systems.
  - Example Use Case: Many inventory management systems in the retail sector were built using RPG.

Industries & Their Programming Needs

Different industries had different computing challenges, and programming languages were tailored to address these needs.

Finance & Banking (COBOL)

- Banks needed systems that could handle large-scale transactions with high precision.
- COBOL was optimized for financial records, making it the backbone of banking systems.
- Many legacy COBOL systems are still in use, so banks struggle to modernize their technology.

Scientific Research & Aerospace (FORTRAN)

- NASA, the military, and research institutions required complex mathematical calculations for physics simulations, orbital mechanics, and aerodynamics.
- FORTRAN was designed to handle scientific workloads efficiently and was the primary programming language for engineering models and supercomputing applications.

Government & Military Systems (COBOL & FORTRAN)

- Social Security, IRS, and defense systems were primarily written in COBOL for financial records and FORTRAN for simulations and weapons calculations.
- The Department of Defense (DoD) later developed ADA, a language meant for mission-critical software in the military and aviation industry.

Retail & Manufacturing (RPG & COBOL)

- Large retailers used IBM mainframes running COBOL and RPG to manage inventory, sales, and logistics.
- Even today, many supply chain and point-of-sale (POS) systems still rely on COBOL-based applications.

Working on the Weekends – No Remote Access

One of the most significant differences between IT work in 1983 and today was the lack of remote access.

- If the mainframe needed maintenance, operators and programmers had to come into the office physically.
- There was no VPN, cloud access, or remote troubleshooting.
- If you were on call for a weekend issue, it meant driving into the data center, manually mounting tapes, and running diagnostics in person.
- The Halon fire suppression system in mainframe rooms meant that if a fire started, you had 30

seconds to get out before the system released gas to suffocate the fire—and you, along with it, if you didn't leave in time.

Unlike today, when engineers log in from home to restart servers, back then, you had no choice but to be on-site for any IT maintenance and emergencies.

Programming Legacy – What's Still Here Today?

Even though many of the programming languages of the 1980s have faded into obscurity, their impact is still felt today.

1980s Language	Modern Equivalent	Legacy Impact
COBOL	Still used in finance & government	Runs legacy banking & payroll systems
FORTRAN	Still used in science & engineering	Used in physics simulations, NASA, and weather forecasting
Pascal	Evolved into Object-Oriented Programming	Influenced languages like Delphi, Java, and Python
C	Still widely used	The foundation for C++, Java, Python, and most modern languages
RPG	Used in IBM AS/400 systems	Still found in legacy business applications

The Programming World Before Personal Computers

Programming in the 1980s was not easy. It was a highly specialized field requiring deep knowledge of hardware and software.

Unlike today, where almost anyone can learn Python or JavaScript in a few weeks, coding requires a deep understanding of system architecture, memory management, and data storage. If you made a mistake, there were no quick fixes; you had to rewrite entire programs, resubmit jobs, and wait hours or days to see if they worked.

This was before personal computing, agile development, version control, and automated testing. Every line of code had to be perfect, or the entire system could crash.

But despite the challenges, these early programming languages built the foundation of the modern digital world. Without COBOL, FORTRAN, Pascal, and C, we wouldn't have the banking systems, scientific breakthroughs, and technological advancements we rely on today.

Mainframe terminal – Programmers entered code line by line on these monochrome screens, often without real-time feedback.

Punched Cards – The Tedious Way to Code

Before modern programming languages, developers had to write code using punched cards. Each card represented a single line of code, and stacks of these cards made up an entire program.

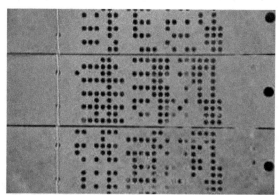

Punched cards *– Each card held a single line of code, physically fed into the mainframe to run programs.*

I still remember receiving boxes of punched cards, sometimes 20,000 or more in a single deck. Each card had to be fed into a machine, which would read the holes punched into it and submit a request to the mainframe.

Challenges of Punched Cards:

- If one card were out of place, the entire program could fail.
- There was no debugging on a screen—you had to check for missing or misplaced cards manually.
- Dropping a deck of punched cards was a disaster—if they got mixed up, it could take hours to reorder them.
- Programmers had to write code with extreme precision, as errors were difficult to correct.

A coworker dropped a box of 20,000 punched cards, completely ruining the sequence. It was impossible to reassemble the cards manually, so the programmer had to re-examine the cards and put them back together or completely re-punch them, wasting more paper.

Debugging Tools & Strategies

In the 1980s, debugging was far from today's step-through debuggers, breakpoints, and console logs. Programmers had to rely on a combination of printed output, system logs, memory dumps, and sheer mental discipline to track down bugs—often without knowing if a program would run successfully until hours after submission.

No Real-Time Feedback

There were no IDEs or interactive debuggers. Once a program was submitted, it ran as a batch job on the mainframe. If the job failed, it would return a printout showing where it broke—but often without much explanation.

The Role of Printouts

Most debugging happened on paper. Developers would receive massive stacks of green bar paper filled with output logs, register contents, memory addresses, and hex dumps. These printouts helped them trace what the program was doing step by step, assuming it ran far enough to generate output.

Memory Dumps & Hex Readouts

If a program crashed, the system could generate a memory dump—a snapshot of what was in memory at the time of the failure. This usually meant pages of hexadecimal code and register values, which programmers had to analyze manually. They'd look at memory offsets, compare expected vs. actual values, and use reference sheets to decode what went wrong.

Common Strategies:

- **Code Review by Hand**: Programmers often debugged by rereading their code line by lines, sometimes with a colleague looking over their shoulder.
- **Instrumenting Code**: Insert temporary DISPLAY or WRITE statements to show variable values at different points in the program, like an old-school console.log ().

- **Logic Tracing**: Carefully following the logic flow with pencil and paper, drawing flowcharts or control flow diagrams.
- **Trial & Error**: Resubmitting slightly altered program versions to isolate a bug, hoping it would produce a different error message or output.

Time Was the Enemy

Every debugging session was high stakes because job turnarounds could take hours (or even a full day). A single typo or logic error could burn an entire day's work. Programmers had to be methodical, patient, and highly detail oriented.

Chapter 3:

Managers, CTOs, and Bureaucracy in the 80s

By the late 1970s, computing had become an institutional fixture—mainframes were humming in government buildings, universities, and large corporations, and COBOL programmers were busy keeping business logic alive. In previous chapters, we explored how these systems operated and how their presence shaped early computing culture. But as these machines grew in power, so too did the bureaucracy around them. IT was no longer a fringe department in the basement; it was now the organization's significant operational and financial arm. This shift brought visibility, red tape, gatekeeping, and an emerging class of managers who didn't come from tech but held the reins.

The 1980s marked the moment when technology decision-making became less about innovation and more about control. This was the decade when career bureaucrats became CIOs, CTOs, and IT directors, often making critical decisions with little technical background or connection to the end users. What began as a push for "governance" quickly became a bottleneck for progress. In many ways, this era's rigidity and risk aversion directly fueled the rise of the personal computer revolution that occurred not because of management but despite it. What follows in this chapter is a look into that strange paradox: how the people in charge of technology often became its most significant obstacle.

Early IT Leadership and Decision-Making

At the time, most CIOs and IT managers came from business backgrounds rather than technical expertise. This led to risk-averse decision-making, where executives hesitated to embrace new technology for fear of failure.

I recall multiple instances where executives rejected PCs to retain outdated mainframes despite growing evidence that personal computing was the future. This hesitation led many organizations to fall behind as the PC revolution accelerated.

This top-down approach to decision-making, rooted in business logic rather than technical or user-centered thinking, created a wide gap between leadership and those using the systems.

The Lack of User Input

One of the most telling consequences of this leadership gap was the near-complete exclusion of users from critical decisions. Another major flaw of early IT governance was the absence of user feedback. And users, the employees who relied on the systems, had no say in how software and hardware were implemented. This disconnect led to frustration, inefficiencies, and resistance to new technology.

I remember working on a system upgrade where end-users weren't consulted until after the deployment. The result? An utterly impractical system forced employees to find workarounds that negated the upgrade's benefits.

Users rarely had a system imposed from the top down.

As an operator, I constantly experienced this disconnect. Systems would be upgraded without any heads-up, and suddenly, we were flying with blind documentation, no support, just a new way of doing things we didn't ask for. We'd huddle with other operators to figure out how to get through the shift.

When people aren't consulted, systems are built in a vacuum. But what made things worse was the more prominent fear looming over the industry—fear of the new.

The Risk-Averse Corporate Mindset

This deep-seated fear shaped a corporate culture that treated innovation as a liability rather than an opportunity. No company has embodied the status quo more than IBM. One of the biggest roadblocks in IT at the time was the fear of new technology. Mainframes were expensive investments, and IT executives did not want to admit that PCs could replace them.

I remember sitting in conversations after hours where IT leaders and managers mocked the idea of personal computers in the office. "Why would an employee need their computer? That's what terminals are for!"

The same resistance happened decades later with cloud computing. Companies that hesitated to move to the cloud fell behind, just like businesses that refused to adopt PCs.

I wasn't in the rooms where these decisions were made but lived with the consequences. We had painfully outdated equipment, batch jobs that took hours, and zero flexibility to make even the most straightforward improvements. It was obvious that better tools were out there but change always seemed to be stalled somewhere higher. Many executives preferred to double down on what was familiar, even when it became clear that the future was shifting. No company has embodied the status quo more than IBM.

IBM's Market Control & The Cost of Mainframes

IBM's overwhelming dominance wasn't just technical but cultural and financial. IBM dominated the computing market, holding an almost unshakable monopoly. Their systems were so integrated into business operations that companies had no choice but to continue paying millions for IBM hardware, software, and support contracts.

To put this in perspective:
- IBM 370 mainframes in the 1970s cost between $200,000 and $4 million.
- IBM 3081 mainframes in the 1980s had a base price of $3.5 million, with operating costs reaching millions annually.
- Software licensing and maintenance contracts could add another $500,000 to $1 million annually.

The high cost of mainframes made it extremely difficult for companies to justify switching to newer technology. They were locked into IBM's ecosystem, often with custom-built software that could not easily be migrated to another system.

The IBM ecosystem locked companies into decades of contracts and custom code

That lock-in wasn't just financial for us operators; it showed how limited we were. Everything was tied to IBM's rules and tooling. I remember asking if we could try something different or automate a task, only to be told, "That's not how it works with IBM." It was like the system ran us—not the other way around. But beyond the cost and contracts, another critical piece often went unnoticed—the experience of the people using these systems daily.

Mainframes vs. PCs in the 80s

Feature	Mainframe	PC
Cost	$3M+	<$5K
User Access	Centralized	Individual
Flexibility	Rigid	Adaptable
IT Control	Total	Minimal

Customer Experience in the Mainframe Era: The Slow & Rigid Systems

In the mainframe era, the concept of "user experience" didn't exist in the way it does today. Technology was built to serve businesses, not individuals, during the mainframe era. Unlike today, where user experience is a priority, back then, systems were designed with efficiency and security in mind, often at the expense of usability. Employees had to adapt their workflows to fit the system rather than the system being built to enhance productivity.

That struck me as an operator. We had to adjust everything we did to fit the system's quirks. Efficiency was not considered unless it was from a system admin's perspective. I often had to take notes on paper to manage job queues and scheduling because the system couldn't give me the real-time information I needed.

One of the biggest frustrations was the batch-processing model. Jobs were queued and executed in order, and they often took hours. This meant that if an error occurred, you wouldn't know until the next day when you received the output. Simple processes that today take seconds—such as generating a report or updating records—were manual, multi-step tasks that could take hours or even days.

One of the most exhausting parts of being an operator was feeding tapes for specific programs. Some jobs, usually ones written by programmers who weren't thinking about operational overhead required hours of nonstop tape mounting. I'd be back and forth to the tape library, loading one reel after another, sometimes wondering if it would ever end. Event on occasion mounting the same tape over and over again.

The system couldn't queue them efficiently, so each tape had to be manually loaded and confirmed before the job could continue. This wasn't rare—it was routine. The process was slow and repetitive, clearly indicating how rigid technology was.

In today's cloud storage and real-time access world, it's almost hard to imagine how dependent we were on physical media just to run a report. Yet, despite these limitations, some industries still managed to innovate in ways that would set the stage for future breakthroughs.

Operator's View: Life Under the Red Tape

As an operator, I didn't make decisions—I responded to them. System changes would appear without warning, documentation, or explanation. We'd show up for our shift and discover a new job stream, a different backup process, or a modified tape schedule. And if it failed? Our job was to figure it out—usually by trial and error.

We weren't asked for input on how things worked, even though we kept the systems running 24/7. The people writing the programs rarely saw what their code did to the workflow on the floor. Some jobs required dozens of tapes—each manually mounted, read, and swapped. I remember nights when I was feeding tapes for hours, barely able to sit down between loads. There were more efficient ways to structure those jobs, but we didn't have the authority to change them. We just kept the wheels turning.

We took pride in what we did, but the bureaucracy meant everything moved slowly—decisions took weeks, and even simple fixes needed layers of approval. There were moments when the system was down, a workaround was evident, and I still had to wait for permission to act.

We kept the systems running but often felt like passengers, not drivers.

Feeding tape after tape for a single job was common for operators
Case Study: Airlines & Reservation Systems

A clear example was the airline industry, which leaned into early computerization to tackle its enormous logistical challenges. Airlines were among the first businesses to rely heavily on mainframes.

In partnership with IBM, American Airlines developed the Sabre (Semi-Automated Business Research Environment) reservation system, one of the earliest large-scale uses of computerized transaction processing. This system allowed travel agents to check seat availability and make reservations electronically, eliminating the need for paper-based tracking.

However, even with automation, these systems were still limited by the technology of the time:

- Reservations were batch-processed overnight, meaning pricing and seat availability updates were not instantaneous.
- Cancellations and modifications were difficult, as databases did not dynamically update in real-time.
- Errors were costly, as agents had to manually verify information before finalizing bookings.

Despite these challenges, systems like Sabre were groundbreaking for their time, laying the foundation for the modern airline reservation platforms we use today. While some organizations focused on practical automation, others reached for sheer computational power to solve scientific and national-scale problems.

Pushing Boundaries: Super Computers

The 1980s saw an explosion of effort to create supercomputers that could break boundaries—technologically, yes, but also economically and competitively. In the 1980s, several supercomputers aimed to push the boundaries of computing power but ultimately failed to impact the market significantly. Here are some notable ones:

CDC Cyber 205 (Control Data Corporation)

- Why didn't it make the mark? CDC was once a leader in supercomputing (with the CDC 6600), but by the 1980s, it struggled against Cray. Cyber 205, launched in 1981, was a vector processor machine overshadowed by the Cray-1 and Cray-2. CDC's internal struggles and declining government contracts led to its downfall.
- Key Issue: It was expensive and didn't offer enough performance advantages over Cray's systems.

ETA-10 (ETA Systems, a subsidiary of CDC)

I saw this supercomputer while working at my government job. It was used for processing, but I was not heavily involved with its operations.

- Why didn't it make the mark? ETA was an attempt by CDC to revive its supercomputing business with a liquid-cooled, high-performance machine that was supposed to compete with Cray. However, ETA couldn't deliver on time, and by the time the ETA-10 came out in the late 80s, it was too costly and underwhelming.
- Key Issue: The CDC shut down the company in 1989 before it could gain traction.

IBM 3090 Vector Processor (IBM)

- Why didn't it make the mark? IBM tried to enter the supercomputer market by adding vector processing to its 3090 mainframes in the late 80s to compete with Cray. However, it was not a dedicated supercomputer, and its performance was weaker than Cray's specialized designs.

- Key Issue: It was marketed as a general-purpose system but couldn't match Cray's raw computing power for scientific applications.

Stardent 3000 (Stardent Computer)

- Why didn't it make the mark? Stardent was formed from the merger of Ardent Computer and Stellar Computer, two companies that tried to create graphics-oriented supercomputers for engineering and visualization. Stardent 3000, released in 1989, had excellent graphics capabilities but struggled to find a market because workstations like Silicon Graphics (SGI) were more cost-effective.
- Key Issue: The market shifted towards workstations, making Stardent obsolete.

Denelcor HEP (Heterogeneous Element Processor)

- Why didn't it make the mark? The Denelcor HEP, launched in 1982, was an experimental parallel-processing supercomputer that tried implementing fine-grain multiprocessing before the industry was ready. It was technically advanced but too difficult to program, and no significant market adoption happened.
- Key Issue: Ahead of its time, software and developer support were lacking.

Intel iPSC (Intel Personal Supercomputer)

- Why didn't it make the mark? In the late 80s, Intel introduced the Intel iPSC (sometimes called the "Cube"), an attempt at a distributed-memory parallel processor. While the iPSC laid the groundwork for future cluster computing, it failed commercially due to high costs and difficulty programming it.
- Key Issue: Clustering was not mainstream, and Intel was not yet a supercomputing leader.

Thinking Machines CM-1 and CM-2 (Thinking Machines Corporation)

- Why didn't it make the mark? These massively parallel supercomputers from the mid-to-late 80s were highly innovative but too complex for most users. They used connectionist AI models but programming them efficiently was a challenge. The company burned through cash and filed for bankruptcy in the 90s.
- Key Issue: To niche and software support lagged hardware capabilities.

ICL DAP (Distributed Array Processor)

- Why it didn't make the mark: A British attempt at massively parallel processing, the ICL DAP (initially designed in the 1970s) was still being marketed in the 80s. However, it suffered from low adoption outside government and research institutions.
- Key Issue: Limited commercial appeal and competitors like Cray dominated the supercomputing space.

NEC SX-2 (NEC, Japan)

- Why didn't it make the mark? NEC tried to compete with Cray by launching the SX-2, a vector supercomputer. While it was one of the fastest machines of the time, US government restrictions on foreign technology hurt its market share.
- Key Issue: Trade restrictions and lack of US adoption limited its success outside Japan.

Why These Supercomputers Failed

Most of these machines failed due to:

- Cray's Dominance – Cray Research dominated supercomputing, leaving little room for competitors.
- Programming Challenges – Many systems require specialized programming, making them difficult to adopt.
- High Costs – Supercomputers were expensive, and only a few organizations (mainly government agencies) could afford them.
- Market Shifts – The late 80s saw a shift toward workstations (like SGI) and parallel computing clusters, which changed the demand for traditional supercomputers.

By the 1990s, many of these companies had gone bankrupt, pivoted, or been acquired as the industry moved toward more scalable and distributed computing models.

Despite their commercial failures, these systems contributed valuable advances in computing architecture, laying the groundwork for modern high-performance computing and cloud clusters.

Another advanced system used for Colleges and Operators was a Virtual Machine (VM). The VM system you worked on in the 1980s was part of IBM's Virtual Machine (VM) operating system, a significant milestone in early computing and networked communication.

This was one of the first large-scale implementations of a "virtualized" computing environment, and it played a crucial role in early internet-style communication between users. As supercomputing hit limits in market appeal, another quieter revolution was unfolding—one that brought people together instead of just crunching numbers.

The IBM VM System and Early Online Communication

Around the same time, systems like IBM's VM began to open doors to communication and collaboration that felt ahead of their time. IBM's VM (Virtual Machine) operating system was initially introduced in the 1970s but became widely used in the 1980s for mainframe computing and remote communications. The system allowed users to run multiple virtual machines on a single mainframe, enabling users (like college operators) to interact in a shared, networked environment.

This environment had a message-passing system that let operators from different locations communicate, essentially an early form of online messaging and email. Typing a message and getting a reply from someone hundreds of miles away, all within seconds, felt like science fiction.

Key Features That Resemble the Early Internet

Here's why this VM system can be considered an early version of the Internet:

- Remote Communication Between Universities and Operators

  - Users could communicate in real-time, like modern instant messaging.
  - College operators like you could connect with other schools and IBM users using IBM's networked infrastructure.

- IBM's PROFS (Professional Office System) and Messaging

  - IBM developed PROFS (an early email and messaging system) that ran on VM.
  - Universities, corporations, and government agencies use it to send messages electronically, like today's email.

- BITNET: A Precursor to the Internet

  - Many university VM systems were connected to BITNET ("Because It's Time Network"), which links IBM mainframes across educational institutions.
  - BITNET supported file transfers, messaging, and email, like the early ARPANET and later the Internet.
  - It allowed universities to collaborate, share resources, and communicate globally.

- VM/370 and VM/CMS

    - You likely worked with VM/370 or VM/CMS, which allowed multiple users to log in to their virtual machines on the same mainframe.
    - This was one of the first multi-user systems where operators could send interactive messages to each other.

- Early Online Chat and Forums

    - Some text-based messaging systems within VM functioned like modern chat applications.
    - Operators and system users could "chat" through direct messages, later evolving into internet-based messaging.

My first exposure (1986) to online communication was before AOL, the web, and widespread email. I experienced what would eventually become the Internet using IBM's VM technology. I went on a blind date with an operator just by talking to them on the computer. Go figure!

Although I had never considered my experience with chat at the time, I remember talking to one operator night after night using email and chat rooms. I went on a blind date, my first experience meeting someone I had never seen in person. There was no way to send a picture beforehand because there were no scanners. So, this was voice-over via phone, chat, and email only.

Again, not knowing what she looked like, I found out she wasn't who she described herself as, and we did not really connect when we met. We still stayed friends online, but my first experience set the tone for dating online forever.

Here's a way to integrate it:

- Personal Experience:

    - I would log into the VM system, send messages to operators at other colleges, and discuss system issues or chat.
    - It was exciting to have a way to chat with other people, but I also didn't realize I was part of a historical shift in technology.

- Comparison to Today:

    - Everything was characters, no pictures or emojis, an early form of instant messaging and email.
    - BITNET and VM communications were the precursors to today's internet-based messaging.

- The Bigger Picture:

    - This era was when mainframes, not personal computers, were communication hubs.
    - As we know, it didn't exist, but these IBM networks paved the way for the Internet.
    - Many tech leaders of the 80s used VM and BITNET as their first online experience.

Chapter 4:

The Start of Change

I still remember the frustration of programmers standing outside the data center, pleading for a report that wouldn't be available for another 24 hours. The mainframe ruled the building—but cracks were forming in its armor.

By the late 1970s and early 1980s, businesses started questioning the inflexibility and high cost of mainframe computing. Employees and department managers wanted more control over their data and workflows, but mainframe administrators resisted changes, fearing security breaches and inefficiencies.

The shift toward personal computing was not just about convenience; it was about empowerment. As technology evolved, users expected faster access, more straightforward interfaces, and real-time data processing. The demand for change would ultimately lead to the rise of personal computers and decentralized computing, completely transforming the IT landscape.

In the 1980s, IBM marketed its mainframes, emphasizing business reliability, performance, and scalability. Here's a breakdown of how IBM positioned its mainframes during that era:

Enterprise Focus: "Nobody Gets Fired for Buying IBM"

IBM was synonymous with trust and reliability in enterprise computing. Its mainframes were marketed as the backbone of business operations, capable of handling mission-critical applications. The famous saying in IT circles, "Nobody gets fired for buying IBM," reflected its dominance and the perception that it was the safest choice for large businesses.

Bundled Solutions & Vertical Integration

IBM didn't just sell mainframes—they sold entire ecosystems:

- Hardware: IBM System/370 and later the IBM 308x and 3090 models
- Software: IBM MVS (Multiple Virtual Storage), VM (Virtual Machine), and later DB2 for database management
- Services: Custom support and consulting for businesses to integrate IBM technology
- Networking: SNA (Systems Network Architecture) for corporate connectivity

IBM positioned its mainframes as an all-in-one solution, ensuring companies stayed locked into IBM's ecosystem.

A single terminal in a mainframe-era computer room—where access meant discipline, and every key press counted.

Leasing Instead of Buying

- IBM used leasing models rather than outright sales. This strategy made it easier for companies to adopt IBM mainframes without substantial upfront costs.
- By leasing, IBM retained control over hardware upgrades and support, keeping customers dependent on their technology.

Performance and Scalability

- IBM marketed its mainframes as scalable and robust, highlighting batch processing, high availability, and multiuser capabilities.
- They emphasized the ability to handle millions of transactions, making them essential for banks, insurance companies, and government institutions.

Competitive Edge Against Mini and Microcomputers

By the late '80s, IBM faced competition from DEC's VAX mini-computers and the rise of microcomputers (PCs). To counter this, IBM:

- Introduced smaller mainframes like the IBM 9370 to target mid-sized businesses.
- Pushed networking solutions (SNA, Token Ring) to show that mainframes could seamlessly connect with smaller computers.
- Emphasized centralized control, arguing that mainframes provided better security and management than decentralized PC networks.

Marketing for IT Executives

IBM targeted CIOs and IT directors rather than individual users, making decisions at the enterprise level. Their pitch was that IBM mainframes provided:
- Reliability: "Zero downtime" computing
- Security: Centralized control over enterprise data
- Longevity: Systems designed for long-term investment.

Leveraging Partnerships and Industry Influence

IBM had strong relationships with banks, airlines, and Fortune 500 companies, ensuring its mainframes were the standard in industries that required high-volume transaction processing. These enterprise-focused strategies weren't just technical; they were also carefully communicated through aggressive and polished marketing.

IBM's Advertising Approach

IBM's marketing included:

- Print Ads in Business Magazines (e.g., Fortune, BusinessWeek)
- Direct Sales with Dedicated IBM Representatives
- Trade Shows and Conferences (IBM showcased its latest mainframes at IT and business expos)

IBM's mainframe marketing in the 1980s was all about positioning itself as the gold standard for enterprise computing, ensuring businesses remained in its ecosystem.

However, by the late 80s and early 90s, the rise of client-server computing and UNIX-based systems challenged IBM's dominance, forcing them to adapt their strategy. While IBM focused on selling reliability to CIOs, a quiet revolution in communication was underway, built on the same mainframe infrastructure.

Modems were used to connect remotely in the 80s

First Internet Sitings

I have had many experiences working with IBM VM; early networking is historically significant. It represents one of the first implementations of virtual computing and online communication, eventually evolving into email, chat systems, and even modern social media.

In the 1980s, modems and networking technologies used to support the IBM VM system relied on mainframe networking protocols, dedicated leased lines, and early packet-switched networks. Here's a breakdown of what supported VM communication during that time:

SNA (Systems Network Architecture) – IBM's Backbone

- IBM's SNA (Systems Network Architecture) was the primary networking protocol allowing the communication of IBM mainframes running VM/370.
- It structured data flow between terminals, remote systems, and mainframes in a hierarchical network.
- It was the enterprise equivalent of what TCP/IP would become.

SNA Components Used for VM Communication:

- VTAM (Virtual Telecommunications Access Method) – Handled mainframe network connections.
- LU (Logical Units) – Allowed terminals and users to interact with the mainframe.
- PU (Physical Units) – Managed network hardware (e.g., controllers, modems).

BITNET – An Early Pre-Internet Network

- BITNET ("Because It's Time Network") was an IBM-based academic network that allowed universities to connect VM systems for email, file transfers, and messaging.
- Used RSCS (Remote Spooling Communications Subsystem) to send messages and files between mainframes.
- Like ARPANET but based on IBM mainframes and SNA.
- Non-interactive: It wasn't real-time like today's internet; messages were sent in batch mode.
- Leased-line modems (usually 9600 baud or lower)
- X.25 packet-switching networks for international links.
- SDLC (Synchronous Data Link Control) for reliable mainframe-to-mainframe communication.

3270 Terminals and Remote Access

- IBM 3270 terminals were widely used to access the VM system remotely.
- They used coaxial cable connections for on-premises access.
- When accessing VM systems from remote locations, users connect through:
  - IBM 3705 Communication Controllers – Managed remote terminal traffic.
  - Dial-up modems (via acoustic couplers or direct lines).

Modems & Dial-Up Access

For remote access and early networking, IBM VM systems used dial-up modems connected to mainframe terminals or controllers

Remote Deployment in Perspective

	Then (Early 1990s)	Now (2020s)
Remote Access	Dial-up modem (physical line sync)	VPN, cloud-based access, remote desktops
Collaboration	Phone calls, handwritten notes	Teams, Zoom, real-time co-authoring
Installation	On-site technician required	Cloud deployment via admin center or script
Risk Factors	Physical travel, unknown security risks	Secure tunnels, MFA, geo-fencing
Support Coordination	Manual sync with internal tech back home	Global support desks, self-healing systems

Types of Modems Used:

- Acoustic Coupler Modems – Early modems where the telephone handset was physically placed into the coupler.
- Hayes Smart modem (Bell 103, 212A Standards) – 300 to 1200 baud in the early 80s, later 2400 and 9600 baud models emerged.
- IBM 3865 Modem – Designed for IBM's SNA and SDLC networks.

Networking Protocols Over Modems:

- X.25 – Early packet-switched protocol used for IBM VM networked systems.
- BSC (Binary Synchronous Communications) – An older IBM protocol for serial communication.
- SDLC (Synchronous Data Link Control) – IBM's standard for networking in SNA environments.

T1 and Leased Line Connections

- Dedicated leased lines (T1, 56K, or lower speeds) were used for permanent connections between mainframe VM systems.
- IBM 3745 Communications Controllers allowed multiple leased-line and dial-up connections to be managed.

Early TCP/IP Adoption for VM

- While SNA was IBM's proprietary network, some universities started implementing TCP/IP on VM systems in the late 80s.
- IBM's TCP/IP for VM allowed VM systems to communicate over early internet protocols, bridging mainframes to the growing internet.
- BITNET gateways started enabling VM users to email ARPANET users.

Your VM System and Early Networking

When you were communicating with other VM operators at universities, your system was connected through:

- BITNET (for university-to-university networking).
- SNA over leased lines or dial-up modems (for internal IBM corporation or university networks).
- X.25 or SDLC for long-distance connections.

This setup was the foundation of the early Internet, making my experience with VM systems a key part of digital history. I was one of the early adopters of networked computing, experiencing real-time mainframe messaging, remote access, and early networked collaboration before the Internet became mainstream.

A Turning Point in Computing

The mainframe era was defined by control, stability, and security, but it lacked the flexibility needed to keep up with growing business demands. Companies that relied solely on centralized computing models struggled to adapt as the world moved toward distributed computing and networked systems.

The next revolution in computing—the personal computer—was inevitable. As more businesses began to embrace decentralized processing, those who resisted would soon find themselves left behind.

I will explore the rise of personal computing, the battle between mainframes and PCs, and the impact of shifting power from IT departments to individual users.

BITNET	1981
IBM 3090	1985
TCP/IP on VM	1987-89

Historic Networking Technologies

The Early Days of My Journey

Looking back at the mainframe era, it's clear that this was the dawn of enterprise computing. It was a time of massive machines, inefficient processes, and rigid corporate structures—but also a time of innovation and learning.

Looking back, my time at the tape library introduced me to the fundamental principles of computing, data management, efficiency, and the importance of precision. It also showed me how technology evolves.

We thought mainframes would last forever, but the personal computer would disrupt everything in just a few years.

My time at the agency's tape library and mainframe computing was just the beginning of my career, but it taught me:

- How to organize and retrieve massive amounts of data manually.
- The importance of efficiency in computing (which became crucial in later roles).
- What emerging technologies—like AI today, start as niche tools before becoming mainstream?
- To embrace change in my role as a technologist
- Understanding how what you learned builds upon each other
- Learned about centralized computing and the risks and benefits

Compared to today, AI again centralizes data, much like mainframes did decades ago. The computing cycle is coming in full circle, and those who don't adapt like IBM's early resistance to PCs may find themselves left behind. Let's look at the comparisons between then and now.

Centralized computing: Two eras driven by data

Back in the mainframe era, everything revolved around centralized control. The data lived in large tape libraries, and running a job meant feeding the mainframe precisely what it needed—sometimes mounting the same tapes multiple times for a single long-running batch job. The mainframe didn't act until a human operator told it to, and even then, everything depended on the availability of the correct data at the right time. Without those tapes, nothing happened. This process, though manual, made it clear: data was the lifeblood of computing.

Today, that principle hasn't changed—only the scale and sophistication have. Modern AI data centers are built on the same foundation: centralized, high-powered computing environments driven entirely by data. Instead of physical tapes, they consume digital information from sensors, applications, users, and historical archives. Once that data is in place, AI systems don't wait for instructions; they analyze, predict, and often act autonomously. Human involvement is no longer required to mount data or kick off the job.

Like the mainframe, these AI systems exist behind a curtain of complexity—most users don't understand how they work, only that they deliver results. Once we trusted mainframes with financial reports and payroll, we now trusted AI with decision-making, diagnostics, and real-time personalization. In both cases, a central authority is lifting the heavy, and an end-user hopes the system gets it right.

Just as mainframes created organizational dependency through reliability and control, AI systems are beginning to do the same—consolidating power in cloud data centers and algorithmic engines. What once required human orchestration now operates independently, but the heartbeat remains the same: data flows in, results come out, and the rest of us adapt around the machine. We've come full circle—from mounting tapes to feeding models—but the principle is unchanged. In computing, data has always been a trustworthy source of power.

In the next chapter, I'll step out of the tape library and mainframe operations and into the world of Personal

Computing, where I'll learn how to navigate tech and accept the challenge of the PC's rise.

Chapter 5:

The Shift to Personal Computers

By the late 1980s, the computing world was on the edge of transformation. The previous chapters have explored the dominance of mainframes, the rise of bureaucratic IT structures, and the tightly controlled environments shaped by IBM and centralized systems. But a quiet disruption was building.

The idea that computing power could be placed on every desk—outside the data center wall was once laughable. Yet, as we moved toward the decade's close, cracks began forming in the mainframe monolith.

PCs were no longer just for hobbyists or secretarial tasks; they were becoming tools for real work. And while IT leadership clung to legacy systems, users and departments were hungry for agility, speed, and control. This chapter marks that turning point—the era when personal computing challenged the old guard and reshaped the business world forever.

Where Was Technology in the Late 1980s?

The world was very different in 1988:

- No PlayStation, no Xbox, no MP3 players.
- The Nintendo Entertainment System (NES) dominated gaming, and Dragon's Lair was introducing animated storytelling to arcades.
- CD-ROMs were appearing, replacing floppy disks for software distribution.
- Blockbuster was the entertainment king, with people renting VHS tapes instead of streaming movies.
- No Wi-Fi, no wireless networking—offices were filled with thick cables running under carpets and along walls.

Meanwhile, the tech giants were battling for control of the future:

- Bill Gates was focused on Windows 3.0 and 95, preparing for a GUI-based world.
- Steve Jobs was ousted from Apple in 1985 and worked on NeXT computers, waiting for his chance to return.
- IBM was still trying to hold onto mainframe dominance but was beginning to lose relevance in the PC space.

The PC industry was about to explode, and I was in the middle of it all, witnessing the transformation firsthand.

1985-1990: PC Curiosity & First Steps

In the late 1980s—1990s, I witnessed firsthand how personal computers (PCs) began sneaking into offices. At first, they were treated as experimental machines for secretaries and accountants rather than IT professionals. The mainframe teams—where I still worked—mocked the idea that a small machine on a desk could replace the mighty IBM and VAX mainframes running entire corporations.

I worked in various industries during this era of the PC revolution, including government, beverages, home mortgages, and as a travel PC tech. PCs were evolving rapidly, gaining more memory and storage capacity. Floppy disks emerged, providing a way to share and transfer files. Though they weren't always dependable, you had to keep a backup on the PC, as necessary.

Looking back, I saw many people laughing at and fighting against PCs. I noticed IBM was fighting because of the Mainframes and holding on to the cost of those mainframe leases and purchases because there was more than one PC. So, the revolution had started.

While working at a large mortgage company in the early 1990s, I witnessed a significant shift with Telex PCs as personal computers began making their way onto office desktops. These machines, equipped with 5 ¼-inch floppy drives, were incredibly slow but represented the first wave of desktop computing in the workplace. Their business applications were limited then, as very little software could fully exploit its capabilities.

Back then, typewriters were still the primary tool for writing letters, which had to be sent through the postal service. With no email or messaging platforms like Exchange, Teams, or Skype, communication depended entirely on physical mail—often taking days to reach its destination. If a letter got lost or delayed, the sender had no choice but to wait or send another, sometimes repeating the same inquiry multiple times. It was a slow, inefficient process that led to frequent miscommunication and wasted time, something today's digital world has largely solved.

However, PCs quickly evolved. Within a few months, IBM PS/2 computers started appearing in offices, offering more memory and storage capacity than their predecessors. Early on, storage was not what it is today, pretty dependable. Using these first computers, files could quickly become corrupted, making it necessary to keep a backup on the PC's hard drive.

At this stage, all PCs relied solely on the command line interface, with no graphical user interface (GUI) in place yet. Users had to manually enter commands to move, delete, or copy files, making even the simplest tasks cumbersome compared to today's intuitive computing environments.

All PCs operated using command-line interfaces, meaning there was no graphical user interface (GUI) yet. If you wanted to move, copy, or delete files, you had to use typed commands; there was no mouse-driven navigation.

We saw a significant change in the number of PCs ordered, and the sweep of PCs was used everywhere but mostly in accounting. We would have a skid of PCs in the office to be set up and configured for delivery. During this time, I was still working as an operator, and two guys were delivering PCs to users' desks at the mortgage company. They would configure the PCs with an MS-DOS command line and take them to the user's desk for software installations and connection to the network.

While working one day, I noticed two coworkers delivering PCs to users' desks. They manually configured them with command-line interfaces before installation. I started helping them unbox and setting up the PCs. Eventually, I became skilled in configuring systems, preparing them for delivery, and establishing network connections.

Eventually, the guys I was working with in Operations got upset and didn't want me to help them anymore. They thought I was doing their job for them. My goal wasn't to do their job for them but to learn what they were doing.

I offered to teach the other operators and invited them to help, but they refused. Instead, they grew increasingly frustrated and resentful. Still, I kept going because I knew it was right for me—I was learning and improving, and they weren't my bosses.

I balanced this with my regular operations job for about three to four months, handling backup tapes and other mainframe-related tasks. However, some of my operations colleagues weren't happy. They accused me of doing someone else's job and even refused to learn from me. Despite the negativity, I kept pushing forward.

The risk paid off. Eventually, I was given a cubicle and a $10,000 raise to oversee PC deployment and network connectivity installs. When our company moved to a new building, I helped move and reinstall the network from the ground up. I ran and labeled all the coax cabling in our new building and successfully migrated over 800 users from one location to another. It wasn't just a raise it felt like a reward and a personal ticket for the future of technology.

During this time, I was still active in the music community in DC. As a musician and a risk-taker, a God given talent (lol) I often found myself drawn to opportunity like a magnet. Some musicians are wired to see possibilities where others see uncertainty. That mindset, the ability to just say *"yes"* when something unexpected happens, took me into rooms and countries I never could've planned for.

At the time, I was playing in a Jimi Hendrix cover band—and not just any cover band. We were a three-piece band that had a sound, a style, and a lead guitarist named Stan Hall, who could not only play Hendrix note-for-note but could do it behind his back. He lived and breathed that sound. He told me of the night he was playing at a club in Georgetown, and something unexpected happened, something that would change the trajectory of his musical journey.

A club owner from Morocco happened to be visiting the U.S. that night. He came to his show and was blown away. After his set, he approached Stan and asked him—without hesitation if we would come to play at his club in Moracco. At first, it sounded unbelievable. I mean, who gets invited to fly across the world just from playing a club set in D.C.?

But it was real. And he said yes.

That moment reminded me that sometimes, doors open not because we knock—but because we show up with excellence and authenticity. It also showed me how God can use any stage, any night, any talent He's given you to position you for more. All you must do is be ready and be bound by Him.

"A man's gift makes room for him and brings him before great men." – Proverbs 18:16

That experience was more than a cool story. It became another layer in my understanding of risk, timing, and trusting the moment. Looking back, it's one more thread in the fabric of a journey that always seemed unpredictable but never without purpose.

That club owner? He was serious. A few weeks after that show in Georgetown, he contacted us to leave for Morocco in early 1993 for a one-month engagement at his venue. Stan was shocked but he needed a bass player because the current bassist could not leave his job for a month to go to Morocco. Stan called me and I took the gig. Then sure enough, the club owner sent over the plane tickets, wired the funds, and made all the arrangements. Just like that, our band was flying across the world to play in a place we'd only heard about in movies.

The club was called Club Fandango, just outside the city of Casablanca. The only time I'd ever heard that name before was in an old Humphrey Bogart film, so I was excited to go—but nothing could have prepared me for what we'd experience there.

The venue itself was upscale. A two-level spot with a dining area upstairs and a music and bar lounge downstairs. People came from all over the region to see our band—locals, tourists, and even soccer players, government officials, and other people of influence. It was surreal. Every night felt like a movie.

But not everything was glitz and glamour.

On one of my days off, I decided to walk about two miles to the only McDonald's in the country. I was craving something American and figured I could make the walk. I passed shaded areas, clusters of trees, and what looked like empty city blocks. And just as I was nearing the restaurant, a group of five young kids, maybe

5 or 6 years old, ran up to me and started checking my pockets.

They weren't violent. They were just desperate.

Thankfully, I had been smart. I had separated my money into two pockets—$20 in smaller bills in one pocket, and another $20 tucked away deeper inside my pants. I let them take what they found without a fight. I didn't yell. I didn't panic. And I didn't retaliate. I was more heartbroken than angry.

Because as I looked around at the homes people lived in, reality hit me hard. There was no middle class. It was either rich or poor—and many of the poor lived in conditions most Americans could not imagine. I saw families staying in partially collapsed buildings, with no roof overhead, yet still hanging curtains in broken windows, trying to make it feel like home. It was sobering.

That experience humbled me in ways I still carry today. I finished my meal at McDonald's, walked back to the place where we were staying, and just sat in silence for a while. I remember from that experience I gave a woman who sat in front of a store begging for money every day we walked by to go to our gig and every coin I got for change while I was in the country I gave to her the night before I left the country. She was so grateful, but I was only giving it to her because I really cared and wanted to bless someone before, I left. Seeing the poverty and all the people who were struggling really impacted me.

That trip reminded me of something I think every American should experience gratitude.

We complain about our problems, but we don't often stop to recognize how much we've been blessed. I've seen what real poverty looks like—and even the poorest person in America has more than many do around the world. You may not have everything you want, but you probably have more than you need. For that, we should be thankful.

"Give thanks in all circumstances; for this is God's will for you in Christ Jesus." – 1 Thessalonians 5:18

That trip wasn't just a gig. It was a life lesson. It was a mirror, showing me the real world beyond American comfort. And it reminded me that no matter where I go, God is always present even in the unexpected places.

1990-1995: Windows 3.0, Office, and Networking

Windows 3.0, released in May 1990, was the first version that genuinely competed with Apple's Macintosh and provided a significant shift from DOS-based computing.

When Microsoft 3.0 was introduced to my company, it sparked a lot of interest due to its connectivity, ability to build networks with the operating system, and ability to transfer files from PC to PC.

Before USB, before plug-and-play—there was the
IBM PS/2. *This machine taught many of us what
computing really meant.*

Key Features That Made Windows 3.0 Revolutionary

1. Graphical User Interface (GUI) – Unlike MS-DOS, Windows 3.0 introduced icons, menus, and multitasking, making it user-friendly.
2. Improved Performance – It supported protected mode memory management, allowing Windows to utilize more than 640KB of RAM, a limitation in MS-DOS.
3. Enhanced Graphics Support – Supported VGA and SVGA resolutions, making Windows visually superior to earlier versions.
4. Introduction of Program Manager & File Manager – Replaced the MS-DOS command line with a point-and-click interface.

5. Better Software Support – Companies began developing Windows applications, moving away from MS-DOS programs.

Sales & Market Impact

- Sales in the First Year: Over 2 million copies, making it the most successful Windows release.
- By 1991, Microsoft became the world's largest software company, surpassing IBM in software sales.
- By 1992, Windows owned over 60% of the desktop market, setting the stage for Windows 3.1 and later Windows 95.

Windows 3.0's Legacy

Windows 3.0 proved that Microsoft could dominate the PC industry, paving the way for Windows 95, which would later revolutionize computing. More than just an upgrade, Windows 3.0 introduced a new way to manage software and interact with the computer—icons, windows, and multitasking replaced the cold command line. It began a more user-friendly era and rapidly shifted how businesses approached technology.

While writing this book, I remembered installing an application called Quattro Pro many times during that time. It was the first spreadsheet software I worked with that felt advanced—it reminded me of what Excel would later become.

When we deployed it, the users' PC often needed extra storage mounted just to save their files. At the same time, many departments, like customer service, still relied on terminals connected to the mainframe, which continued to store and manage all the critical business data.

This contrast between the emerging PC tools and the old terminal systems made me realize something important: the PC was no longer a toy. It was becoming a serious tool, and my thinking about what computing could be was shifting.

I am sure that more companies have started experimenting with PCs and are trying to utilize them for some of their strengths. It's like looking at apps today. Apps had different abilities back then, and hardware had different abilities because the hardware mattered.

Then, out of nowhere, the mortgage company purchased Microsoft Office, and it came on the scene, which made many people happy because they could craft letters and use Excel to crunch numbers and inventory. Presentations also changed, and PowerPoint could now simulate past paper presentations. Old hardware projectors were trashed, and new hardware for projecting from computers changed how you could create and present presentations.

Now, companies could:

- Write documents (Word)
- Use spreadsheets (Excel)
- Create presentations (PowerPoint)

Microsoft Office: The Suite That Dominated Business Computing

Microsoft Office revolutionized productivity by bundling essential applications like Word, Excel, and PowerPoint into a single package, which became the default business standard.

- First Release: August 1, 1989 (for Mac), November 1990 (for Windows)
- Original Applications: Microsoft Word, Excel, and PowerPoint
- Operating System: MS-DOS (early versions), later Windows 3.0/3.1

Why Microsoft Office Became a Game-Changer:

1. Bundled Software Approach—Before Office, businesses had to buy WordPerfect for word processing, Lotus 1-2-3 for spreadsheets, and Harvard Graphics for presentations. Office provided everything in one suite.
2. Tight Windows Integration – Office was optimized for Windows, giving it an advantage over competing software that struggled to transition from MS-DOS.
3. Corporate Licensing – Microsoft aggressively sold Office to businesses in bulk, making it cost-effective for companies to standardize on one suite.
4. Usability & Compatibility—Microsoft ensured that file formats worked seamlessly across applications, making them easier for businesses to adopt.
5. Regular Updates & Expansion – Over time, Microsoft added Outlook (email), Access (databases), and OneNote (digital notebooks).

Sales & Market Impact:

- By 1992, Microsoft Office dominated the software market, overtaking WordPerfect and Lotus.
- By 1995, with the release of Office 95, it had captured 90% of the business market.
- By 2000, Office had become the world's most used productivity suite.
- Microsoft Office contributed billions to Microsoft's revenue, cementing the company's dominance in software.

Today, Microsoft Office 365 (now Microsoft 365) remains the most widely used office suite, generating over $44 billion in revenue annually.

During this time, software debates were common. We were fiercely loyal to Quattro Pro, while others favored Microsoft Excel. One of the most significant limitations back then was that the two programs used completely different, proprietary file extensions Quattro Pro used formats like. wb1 and. qpw, while Excel used .xls.

There was no easy way to open or share files between them, which forced organizations to choose one standard. Ultimately, many adopted Microsoft Office as the official platform, driven by the need for consistency and compatibility.

Looking back, this reflects a recurring pattern I've seen throughout my career: users often develop strong preferences for specific tools, sometimes based on minor features. But in the end, it's rarely the feature that defines longevity; it's the ecosystem and its ability to adapt.

Microsoft won this battle for the desktop, and Quatro Pro and Lotus Notes 123 eventually went out of business. The big reason is that Microsoft had an operating system that supported Office. We occasionally got errors from Quattro Pro and Lotus Notes, so keeping the desktop consistent at that time was very important because of the support.

PC support had just become a booming business back then. Repair shops everywhere want to help you fix hardware and software issues. After this was in place, we saw more malware and viruses popping up, which brought a new Anti-Virus software business, which was a massive change in support for PCs.

The Evolution of PC Connectivity and Expansion

One of the most significant differences in personal computing back then was the absence of modern conveniences like USB or USB-C ports. Instead, we relied on parallel and serial ports such as Parallel, COM1, and COM2 to connect devices like printers and modems. These connections were notoriously finicky, often requiring patience, manual driver installations, and a fair amount of luck to get them working correctly. Plug-and-play was still years away, so every device setup felt like a custom project.

As storage demands grew, SCSI (Small Computer System Interface) drives emerged as a game changer. They allowed for significantly faster data transfer between devices and supported daisy-chaining multiple peripherals to a single controller. This made sharing large files between PCs or running external hard drives and CD-ROM towers possible—an essential step for media professionals and early enterprise users. However, SCSI wasn't without its quirks; it often required careful configuration, and drive failures could mean complete system reboots.

PCs began incorporating expansion slots like ISA, EISA, and eventually PCI (Peripheral Component Interconnect) to support devices like SCSI drives and network cards. PCI became standard in the 1990s due to its speed and flexibility, allowing users to easily install everything from network cards to video accelerators and sound cards.

This expandability transformed the PC into a modular platform capable of adapting to professional and personal needs. These slots also paved the way for connecting early projectors, which allowed business professionals to give live presentations directly from their PCs, a giant leap forward in productivity.

Together, these developments laid the foundation for the highly connected, high-performance systems we now take for granted. What once required special cabling, technical knowledge, and hardware tweaking is now solved with a simple USB plug or wireless connection.

Decentralization Begins – Companies Giving Employees Their Computers

The transition from mainframe terminals to personal computers (PCs) in the 1980s was a significant shift, especially for departments like finance, which needed spreadsheets and accounting software that worked better on local machines rather than centralized mainframes. During this period, 5 1/4-inch floppy disks were the primary storage medium before the industry moved to 3 1/2-inch floppy disks and early hard drives.

How Storage Changed in the 1980s and 1990s

PC storage evolved quickly:

- Early 1980s: 5 1/4-inch floppy disks (360 KB - 1.2 MB)
- Mid-1980s: 3 1/2-inch floppy disks (720 KB - 1.44 MB)
- Late 1980s: Hard drives (10-40 MB standard, some up to 100 MB)

Here's a look at some of the most notable PC and Mac models from the 1980s, including their specs and key use cases. This led to where we are today:

Early IBM-Compatible PCs (1980s)

IBM set the standard for business computing, and many finance departments were among the first to receive PCs to replace terminals. PC Models included:

IBM 5150 (IBM PC) – 1981

- Processor: Intel 8088, 4.77 MHz
- RAM: 16 KB (expandable to 256 KB)
- Storage: Two 5 1/4-inch floppy drives, no hard drive initially
- OS: PC-DOS 1.0 (later MS-DOS)
- Why It Mattered: The first IBM PC set the standard for business computing. Used for Lotus 1-2-3, an early dominant spreadsheet application for finance.

IBM 5160 (IBM XT) – 1983

- o Processor: Intel 8088, 4.77 MHz
- o RAM: 128 KB (expandable to 640 KB)
- o Storage: 10 MB hard drive (first IBM PC with a hard disk) + 5 1/4-inch floppy drive
- o OS: MS-DOS 2.0
- o Why It Mattered: Allowed businesses to store more data locally and run financial applications faster.

IBM 5170 (IBM AT) – 1984

- o Processor: Intel 80286, 6 MHz
- o RAM: 256 KB (expandable to 16 MB)
- o Storage: 20 MB hard drive + 5 1/4-inch floppy drive
- o OS: MS-DOS 3.0
- o Why It Mattered: It was one of the first PCs with a built-in hard drive, making it ideal for financial modeling and accounting software.

Other Popular IBM-Compatible PCs

Since IBM set the business computing standard, clones (third-party IBM-compatible computers) emerged, offering similar specs at a lower cost.

Compaq Portable (1983)

- • Processor: Intel 8088, 4.77 MHz
- • RAM: 128 KB
- • Storage: Dual 5 1/4-inch floppy drives; later models had a hard drive

- Why It Mattered: The first IBM-compatible portable computer, popular with accountants and executives who needed mobility.

Tandy 1000 (1984)

- Processor: Intel 8088, 4.77 MHz
- RAM: 128 KB (expandable to 640 KB)
- Storage: 5 1/4-inch floppy drive
- Why It Mattered: A lower-cost IBM-compatible PC, often used by small businesses.

TRS-80: *One of the machines that sparked a revolution. No cloud, no apps—just raw computing.*

Hewlett-Packard Vectra (1985)

- Processor: Intel 80286, 6 MHz
- RAM: 256 KB

- Storage: 10-20 MB hard drive
- Why It Mattered: A business-oriented PC competing with IBM.

Early Apple / Mac Computers (1980s)

While IBM and clones dominated business computing, Apple was focused on creative professionals and education.

Apple II (1977-1985) – Used in Finance, too!

- Processor: MOS 6502, 1 MHz
- RAM: 4 KB (expandable to 64 KB)
- Storage: Cassette, later 5 1/4-inch floppy drive
- Why It Mattered: Some early finance departments used VisiCalc, the first spreadsheet program, which ran on Apple II before moving to IBM PCs.

Apple Lisa (1983)

- Processor: Motorola 68000, 5 MHz
- RAM: 1 MB
- Storage: 5 MB hard drive
- Why It Mattered: It was the First GUI-based computer for businesses, but it was too expensive for finance ($9,995).

Apple Macintosh (1984)

- Processor: Motorola 68000, 8 MHz
- RAM: 128 KB
- Storage: One 3 1/2-inch floppy drive (no hard drive at lunch)

- Why It Mattered: The first successful GUI computer was not widely used in finance due to limited software.

Macintosh Plus (1986)

- Processor: Motorola 68000, 8 MHz
- RAM: 1 MB (expandable to 4 MB)
- Storage: 800 KB 3 1/2-inch floppy drive, supported external hard drives
- Why It Mattered: A step toward business adoption, though still more common in graphics and publishing than finance.

Key Takeaways: The First PCs in Business

- Finance Departments Adopted PCs First – They needed spreadsheets (VisiCalc, Lotus 1-2-3, and later, Excel).
- IBM PCs Dominated Business Use – The IBM PC, XT, and AT were the most common business machines.
- Apple was a Niche Player – Mac was used in design and education, while Apple II was an early spreadsheet pioneer.
- Storage and Power Improved Rapidly – The 1980s saw floppy disks shrink, hard drives grow, and PC speeds increase.

Around this time, Businesses still relied heavily on mainframe terminals for handling customer service and data processing, but this was about to change.

Companies began to experiment with PCs in business operations, recognizing their strengths. As mobile apps today provide different capabilities, hardware capabilities dictate how computers can be used. This era marked the beginning of businesses using customized software applications to gain efficiency.

The finance industry was one of the first to transition from mainframes to PCs, mainly because of the growing need for real-time calculations, spreadsheets, and local financial modeling. With the rise of IBM-compatible PCs, finance professionals and accountants gained access to faster, more flexible budgeting, forecasting, and transaction management tools.

Here's a deep dive into the finance software that shaped the 1990s, from spreadsheet pioneers to accounting and banking systems.

The PC Boom – Who Adopted First?

PC adoption wasn't uniform, some industries embraced it immediately, while others held on to mainframes for decades.

I tried convincing my manager that PC-based networks were the future, but they refused to listen, believing that mainframes were more secure and would stand the test of time.

Industries that transitioned first:

- Retail – Needed better point-of-sale systems and inventory tracking.
- Manufacturing – Used PCs for automation and real-time production tracking.
- Finance & Investment – Needed faster stock trading and real-time analytics.

Industries that resisted change:

- Banking – Stuck with COBOL-based mainframes, many of which still exist today.
- Government & Defense – Legacy systems kept running due to security concerns.
- Healthcare – Hospitals relied on mainframes for patient records for many years.

Industries that shifted away from mainframes first typically valued agility, cost efficiency, and decentralized computing over the centralized control that mainframes provided. The transition happened gradually, with different industries adopting client-server computing, PCs, and cloud-based solutions at various paces. Here's a breakdown of the early adopters that moved away from mainframes first:

Retail and E-Commerce

- Retail required faster, more flexible computing systems to manage point-of-sale (POS) transactions, inventory, and customer data across multiple locations.
- Shift:
    - Early adopters of PC-based POS systems (IBM 4680 series, NCR registers) in the late 1980s and early 1990s.
    - Adoption of client-server databases (like Oracle and SQL Server) instead of relying on centralized mainframes.
    - By the late 90s, companies like Amazon helped push cloud-based computing for e-commerce, eliminating the need for on-premises mainframes.

Manufacturing

- Why they moved first: Manufacturers needed real-time data tracking, automation, and decentralized access across supply chains.
- Shift:
    - Moved from mainframe-based inventory and ERP (Enterprise Resource Planning) systems to PC-based systems like SAP and Oracle ERP.
    - Implemented SCADA (Supervisory Control and Data Acquisition) and PLC (Programmable Logic Controllers) for automation.
    - The adoption of distributed computing allowed factories to process orders, track production, and manage logistics more efficiently.

Travel and Airlines

- Why they moved first: Airline reservation systems were some of the first mainframe-driven applications, but by the 1990s, airlines needed real-time ticketing and decentralized access for customers.
- Shift:
    - Moved from centralized mainframe reservation systems (like Sabre and Amadeus) to distributed cloud-based booking engines.
    - Allow airline partners, travel agencies, and customers to book tickets online, reducing dependency on closed mainframe networks.

Financial Services (But Not Banks at First)

- Why they moved first: Investment firms and stock exchanges required real-time data and high-speed transactions, which mainframes couldn't manage efficiently.
- Shift:
  - Trading platforms moved to Unix-based and later cloud-based systems.
  - High-frequency trading firms adopted PC clusters, supercomputers, and later, cloud services.
  - Banks, however, were slow to transition, as they relied on mainframes for secure transactions and account management (many still do today).

Telecommunications

Why they moved first: The telecom industry requires fast, scalable systems for billing, call routing, and customer service.

- Adopted Unix-based servers and later cloud-based solutions for customer billing and call management.
- Early adopters of distributed computing in the 1990s to reduce mainframe dependency.
- The rise of VoIP (Voice over IP) and digital networks meant less reliance on legacy mainframe-based switching systems.

Healthcare & Insurance

Why they moved first: Hospitals and insurance companies needed decentralized patient records and billing systems.

- Transition from mainframe-based patient management systems to client-server models (early electronic medical records systems).
- Healthcare billing moved to decentralized systems, allowing easier access across different healthcare providers.
- Insurance companies started using AI-driven and cloud-based risk assessment models instead of mainframe batch processing.

Industries That Were Slow to Move Away from Mainframes

These industries held on to mainframes longer due to security, transaction integrity, and legacy dependencies:

- Banking & Financial Institutions – Some still use COBOL-based mainframe systems today for secure transactions.
- Government & Defense – Mainframes are still used in classified systems and legacy applications.
- Utility Companies – Power grids and SCADA systems relied on mainframe-based control systems.
- Railroads & Transportation – Legacy ticketing and logistics systems were slowly transitioning.

The first industries to move away from mainframes were those that required:

- Faster, real-time data access (e.g., stock trading, telecom).
- Cost efficiency and decentralized operations (e.g., retail, manufacturing).
- Scalability for customer transactions (e.g., travel, insurance).

However, mainframes are still used in finance, government, and large-scale data processing. The fundamental shift came when cloud computing became viable, allowing even slow adopters to replace their legacy systems.

The Contrast Between Fast & Slow Adopters – Success Stories vs. Failures

While Apple computers existed, they were rarely used in corporate settings. Apple products were considered more suited for personal use and incompatible with enterprise networks. Their protocols were non-standard, and the mouse-driven interface was seen as a novelty rather than a necessity.

Case Study: Banking & PCs – How Finance Resisted Decentralization at First

As more PCs entered the enterprise space, network management became critical. We used Novell NetWare, the best network management solution at the time. It was complex but provided centralized control over PCs and users.

Customer Frustrations & IT Challenges

The Growing Pain of IT Departments – Managing Decentralized Systems

During this period, IT managers had their visions for technology, often based on what they were already familiar with rather than what was best for the company. I witnessed companies making poor technological choices, leading to lost functionality and unnecessary training costs.

The Explosion of Software Needs – Why Businesses Needed More IT Support

Slow adopters found out the hard way that businesses wait for no one. The rise of PCs revolutionized:

- Communication (Email)
- Collaboration (File Sharing, Networked Devices)
- Data Analysis (Excel, Databases)

Executives become mobile with laptops, eliminating the need for paper reports. Meanwhile, those who stuck with mainframes struggled to compete with the speed and flexibility of PCs.

The PC Revolution Had Arrived

By the mid to late 90s, it was clear that PCs had won the war.

- Windows 95 changed everything.
- Networking & the internet have transformed how businesses operate.
- Mainframes were fading away, except in banking and government.

The next frontier was the internet, and that would change everything again.

The bridge between DOS and the modern desktop—
Windows 95 *changed how the world worked, clicked, and started.*

Job Market Analysis 1994

In 1994, the U.S. labor market experienced significant growth, adding approximately 2.3 million jobs. The employment-population ratio increased from an average of 62.5% in the first quarter to 63.0% in the fourth quarter. **IPUMS CPS**

The technology sector, particularly computer and data processing services, played a substantial role in this expansion. This high-paying industry added 105,000 jobs in 1994, reflecting the increasing integration of technology across various sectors. **Bureau of Labor Statistics**

During the early to mid-1990s, the tech industry underwent a notable shift. Until 1996, most tech employment was in manufacturing, accounting for approximately 60% of the high-tech workforce.

However, service firms began to dominate the tech economy, with 80% of tech workers employed in services by the late 1990s. The computer systems design industry became a significant employee, reflecting the growing demand for software development and IT consulting services. **Federal Reserve Bank of St. Louis**

Overall, the 1990s saw substantial job growth in the U.S., with the total number of jobs increasing by 28 million from 1990 to 2000. More than half of this growth (14.6 million jobs) occurred in the services sector, highlighting the economic shift toward service-oriented industries. **Indiana Business Research Center**

These trends underscore the dynamic nature of the labor market during that period, characterized by rapid technological advancements and a transition from manufacturing to service-based employment within the tech sector.

Servers and LAN Technologies

We then saw two networking frameworks emerge: Ring and Token Ring. The mortgage company I worked for then wanted to use Token Ring, an IBM product, go figure (mainframe). The token ring was new and only 4 MB per second. This was short-lived as new players came onto the scene.

Token Ring Networks: IBM's Networking Technology

Before Ethernet became dominant, IBM introduced Token Ring in 1984 as an alternative local area network (LAN) technology.

- How It Worked: Data was transmitted in a ring topology, where a token was passed to grant permission to send data.
- Speed: It started at 4 Mbps and was later upgraded to 16 Mbps.
- Used In: Large enterprises, especially those using IBM mainframes.
- Advantages: More reliable than early Ethernet because it avoided collisions.
- Downside: Expensive, complex, and slower compared to Ethernet advancements.
- Decline: By the late 1990s, Ethernet (10/100 Mbps) had taken over due to lower cost and better performance.

IBM discontinued the Token Ring in the early 2000s, and it faded into history.

There were also other networks available as well:

Ring Networks vs. Star Networks vs. Ethernet

In the 1980s and 1990s, multiple LAN topologies were competing for dominance:

Ring Network (Token Ring)

- IBMs closed-loop system, where devices were connected in a ring.
- Stable but expensive and complicated software to scale.

Star Network

- Each device is connected to a central hub or switch.
- More fault-tolerant than Ring because if one device fails, the rest stay online.

Ethernet (Dominant Standard)

- Developed by Xerox in the 1970s, commercialized in the 1980s.
- Initially 10 Mbps, later 100 Mbps (Fast Ethernet), then 1 Gbps+.
- Became the default networking standard by the late 1990s, replacing both Ring and Star in most environments.

By 2000, Ethernet was the universal standard for networking.

What Happened to the Operating System That Launched Microsoft

Before Windows, there was MS-DOS (Microsoft Disk Operating System), which was the foundation of PC computing in the 1980s.

- First Released: 1981, after Microsoft licensed QDOS (Quick and Dirty Operating System) from Seattle Computer Products.
- Command-Line Interface: No graphical environment—users had to type text commands to navigate and execute programs.
- Key Versions:
    - MS-DOS 3.3 (1987) – Introduced complex drive support over 32MB.
    - MS-DOS 5.0 (1991) – Improved memory management and added EDIT.EXE (text editor).
    - MS-DOS 6.22 (1994) was the last standalone version before Windows 95 took over.

Why MS-DOS Was Replaced

- Graphical Interfaces (Windows) – People wanted a mouse-driven OS, leading to Windows 3.1 and Windows 95.
- Limited Memory Management – MS-DOS struggled with modern applications needing more RAM.
- No Networking Capabilities – Unlike Novell NetWare or Windows NT, MS-DOS had no built-in networking.

By the late 1990s, MS-DOS was phased out, but its command-line environment (CMD/PowerShell) still exists in Windows today.

The original save icon—back when storage meant 1.44 MB at a time.

IRMA Card: Connecting PCs to Mainframes

Before Windows and networking took over, businesses relied on IRMA (IBM Remote Management Adapter) cards to connect personal computers to mainframes without using dumb terminals.

- Purpose: Allowed IBM PC users to connect to mainframes running 3270 terminal emulations.
- Used In: Banking, finance, and government offices in the 1980s and early 1990s.
- Replaced by TCP/IP networking, Windows-based terminal emulators, and direct client-server applications.

The IRMA card played a crucial role in bridging the gap between mainframes and the PC revolution but was phased out as networking technology improved.

After moving everyone to our new location, we saw PC cards being purchased to connect to the mainframe. The cards were inserted into the PC, and a coax cable was used to connect to the mainframe. This allows PC users to use a PC and connect to the mainframe when needed.

This game changer sealed the deal for PCs' existence today. Most technologies allowing users to connect to an external data source or other shared hardware bring new functionality to the user, further pushing the agenda for leaving the old behind.

We started seeing some Apple products as things progressed, but not many. I must say that back then, Apple was made for use at home or by individual music composers. They didn't work well on networks and were very slow. Protocols were not standard, and many differences existed in using them.

IBM OS/2 – The Operating System That Should Have Won

IBM's short-lived OS2 operating system was also part of my journey. We installed and evaluated it as a server in our network. We randomly scrapped the whole project after many weeks of testing and installation. We wasted money and time on something where Novell was doing a decent job.

IBM OS/2 was one of the most advanced operating systems of its time, and it could have been what Windows became. However, IBM's poor decisions and Microsoft's aggressive strategy led to its downfall.

The Origins of OS/2

- IBM and Microsoft co-developed OS/2 in the mid-1980s to replace DOS.
- OS/2 was designed to be more stable, support multitasking, and handle enterprise workloads better than MS-DOS.
- IBM and Microsoft marketed OS/2 together as the future of PC operating systems.

What Made OS/2 Special?

- True 32-bit multitasking (before Windows 95!)
  - OS/2 had real multitasking long before Windows could do it properly.
  - Windows 3.x was still running on top of DOS, but OS/2 was a fully independent OS.
- Superior Memory Management
  - OS/2 had protected memory, meaning one crash program wouldn't take down the entire system.
  - Windows 3.1 still had major stability issues, while OS/2 was rock solid.
- Better Networking and Security
  - OS/2 supported high-end networking and ran IBM's enterprise software better than Windows.
  - Banks, ATMs, and critical systems used OS/2 for years after IBM stopped supporting it.

How Microsoft Destroyed OS/2

The Microsoft Betrayal (1990)

- While co-developing OS/2 with IBM, Microsoft secretly built Windows 3.0, which was cheaper and had more software support.
- When Windows 3.1 became popular, Microsoft abandoned OS/2 and told developers to write software for Windows instead.
- IBM was left to finish OS/2 alone, but Microsoft had already captured the market.

OS/2 Warp – The Last Attempt (1994)

- IBM released OS/2 Warp, which was faster and more powerful than Windows 95.
- It could run Windows programs better than Windows and had built-in Internet support before Microsoft.
- But it was too late—Windows had already won the marketing battle.

IBM's Big Mistakes

- OS/2 was expensive, while Windows 95 was cheap and pre-installed on PCs.
- Lack of developer support—Since Microsoft encouraged developers to focus on Windows, OS/2 never had enough software.
- IBM only marketed OS/2 to businesses, while Microsoft made Windows the default choice for home users.

The Aftermath

- By 1996, OS/2 was dying, and IBM officially ended support in 2006.
- Some ATMs and banking systems still ran OS/2 into the 2010s because of its reliability.
- But for the public, OS/2 vanished, and Windows became the dominant OS.

The Legacy of Novell and OS/2

- Novell pioneered networking but lost to Microsoft because of slow adaptation and strategic mistakes.
- OS/2 was superior to Windows, but IBM's lousy marketing and Microsoft's aggressive tactics killed it.
- Microsoft bundled everything into Windows Server and Windows NT, wiping OS/2 and NetWare off the map.
- By 2000, Windows was the king of enterprise networking and operating systems.

The Rise of Data Communications – A Catalyst for the Internet Age

As technology advanced, data communications became the foundation for eventually evolving into internet access points and global connectivity. In the early days, businesses relied on modems—one-to-one connections that allowed companies to share customer account information, internal databases, and inter-company communications. However, as demand for faster, more reliable, and scalable solutions increased, businesses began implementing multiplexers (MUXs) and network hubs to facilitate multi-user access across organizations.

My next stop in IT was as a Data Communication Technician. This role exposed me to the complex infrastructure that kept companies connected before the widespread adoption of the Internet. The company I worked for used a Time plex MUX for communications, a highly sophisticated system that allowed multiple data streams to be combined and transmitted over a single communication line. The intricate setup involved dozens of modems, leased lines, and dedicated circuits, all working together to ensure seamless connectivity between partners, clients, and other businesses relying on our services.

Installing and maintaining these connections was one of the most critical aspects of my job. I frequently worked with AT&T, local carriers, and enterprise clients, traveling to customer locations to configure and establish new connections. These installations required precise coordination, as downtime could impact businesses relying on our network for critical financial transactions, customer records, and corporate communications.

I remember setting up this system we used called Infobot; a cutting-edge solution at the time that provided customers with account information through an automated phone system. Callers could dial in, enter their loan number, and receive real-time updates on their mortgage status, payment details, or escrow information. It wasn't flashy, but it worked, and it was one of the first times I saw automation reduce the load on human operators.

Looking back, Infobot was a precursor to the voice-enabled AI systems we see today. In some ways, it reminds me of how modern AI is now being trained to understand natural language and replicate human voices—offering customer support, reading scripts, or even voicing brand advertisements. What seemed like science fiction back then is now the new normal.

You can see how seriously the company took reliability by looking at its use of POTS (Plain Old Telephone Service) lines, along with 56K and T1 modems across the enterprise network. Systems like Infobot were designed with this infrastructure in mind—capable of handling up to eight phone lines simultaneously. Each line had its own number but was centrally connected to an 800 number, allowing multiple customers to call in at once to check their account status.

Network reliability was not optional, it was critical. But unlike today's software-defined, cloud-managed networks, troubleshooting back then was completely manual and highly time-consuming. Diagnosing an outage meant physically inspecting modem banks, cables, and multiplexers. Restoring services often required long hours in the data center, coordinating with telecom engineers, and testing each connection until the issue was found and resolved.

This era of early data communications laid the groundwork for modern networking and internet infrastructure. It was a time of rapid innovation, where companies began moving from isolated systems to interconnected networks, setting the stage for the explosion of enterprise networking, the rise of the World Wide Web, and the birth of the modern Internet.

Timeplex MUX: A Legacy of Early Multiplexing Technology

Timeplex was a major player in the networking and telecommunications industry during the 1980s and early 1990s. It specializes in multiplexing technology, which combines multiple signals over a single transmission line to maximize efficiency.

You weren't online until you heard this thing scream at you

What Was Timeplex MUX?

The Time plex MUX (Multiplexer) was a device that consolidated multiple voice, data, and video signals onto a single network link, reducing the number of required transmission lines and increasing efficiency. It was widely used in enterprise networks, banks, government agencies, and telecom companies.

How It Worked:

- Timeplex MUX allowed multiple data streams to share a single high-speed connection, reducing costs for organizations with multiple remote offices.
- It supported synchronous and asynchronous data for applications like mainframe terminal connectivity, early VoIP, and financial transaction processing.

- It worked on T1, E1, and later fiber-optic networks, helping organizations transition from analog to digital networking.

Industries That Used Timeplex MUX:

- Banking & Finance – Used for secure transaction processing and real-time stock trading networks.
- Government & Military – Employed for classified data communications and early networking of secure sites.
- Telecommunications – Enabled efficient voice and data routing for early telecom networks.
- Airline & Transportation – Used in airport communication systems and early reservation networks.

Why Timeplex MUX Was Popular

- Cost Savings – Combining multiple streams reduced the need for numerous leased lines.
- Reliability – It is known for its stable network architecture, making it a preferred choice for financial institutions.
- Scalability – Allowed companies to scale their network without significant infrastructure changes.

What Happened to Timeplex?

By the mid-to-late 1990s, packet-switched networks, IP-based communication, and Ethernet overtook traditional TDM (Time Division Multiplexing) networks. Companies migrated to TCP/IP-based networking, reducing the need for Timeplex's legacy multiplexing solutions.

- Timeplex was acquired by Unisys in 1987.
- In the late 1990s, Ascom (a Swiss telecommunications company) acquired Timeplex but struggled to keep it competitive.
- Eventually, Timeplex was phased out as companies moved toward modern routers, switches, and IP-based networking.

Legacy & Impact

While Timeplex MUX is no longer in use, its multiplexing principles remain critical in modern networking technologies, including MPLS (Multiprotocol Label Switching), SD-WAN (Software-Defined Networking), and Cloud Networking.

Our mortgage company's data center was massive, filled with rows of racks supporting an array of modems to handle customer connections. As networking technology advanced, particularly with the introduction of Novell NetWare, some of these services began migrating to PCs, allowing for greater automation, and providing portal interfaces for customers and partners. This shift reduced reliance on mainframes and set the stage for a more distributed computing environment.

As mainframes were phased out, our company underwent a significant layoff. Many IT employees were let go, due to the consolidation of services onto servers and the increasing independence of employees using personal computers.

The traditional computer room footprint was shrinking, replaced by server-based infrastructure that required fewer dedicated operators and technicians. Cubicles were constructed in the vacant space for those who worked closely with systems and building maintenance technicians.

While I wasn't directly involved in sales, I saw firsthand how difficult it was for certain technology vendors to gain a foothold in our company. IT managers had their visions of which technologies to adopt, often based on what they were already familiar with rather than the best solutions available. This reluctance to explore new options stifled growth and limited the company's ability to innovate. Instead of selecting cutting-edge technology, management often gets stuck with legacy systems, resulting in reduced functionality, increased training costs, and inefficient for the future. Watching better products get overlooked simply because decision-makers were comfortable with the status quo was frustrating.

As PCs became more widespread, they were used for more than just business tasks. Employees increasingly used them for personal storage, often filling hard drives with non-work-related content. At one company where I worked as a PC technician, I remember discovering 20GB of illegal content on an employee's machine. He was immediately fired, but this was just one extreme example of IT teams' challenges in enforcing computer policies, as personal computing blurred the lines between work and private use.

Slow adopters quickly realized that business does not wait for anyone. Companies that embraced PCs and modern networking gained a competitive advantage, leveraging faster communications (email), networked file sharing, and batch processing for printing. These capabilities revolutionized workflow efficiency and collaboration, making businesses far more productive.

On the other hand, those who remained tied to mainframes struggled to keep up. While mainframes were once essential, they couldn't match the speed and flexibility of PCs. Tasks that took hours or overnight batch processing were completed in real time or within minutes. Previously a slow and cumbersome process, data analysis was now done on Excel and other desktop tools, allowing executives to make faster, data-driven decisions.

The introduction of laptops further transformed business. Executives were no longer tethered to their offices, taking their presentations on the road rather than printing out stacks of paper reports. This shift enhanced mobility and drastically reduced paper waste, signaling the beginning of a digital-first approach to corporate communication.

The Rise in Finance Software

Before PCs, financial calculations were done manually or on mainframes, requiring IT professionals to run batch jobs. The arrival of spreadsheet software changed everything.

VisiCalc (1979) – The First Spreadsheet Software

- Platform: Apple II (later ported to IBM PC)
- Why It Mattered:
    - First software to automate financial calculations
    - Allowed accountants to edit numbers and see changes instantly
    - Used for budgeting, financial forecasting, and balance sheets
    - Finance departments adopted Apple II computers just for VisiCalc before switching to IBM PCs

VisiCalc was the reason many financial professionals first embraced personal computers. It saved hours on manual calculations and reduced reliance on IT departments for economic reports.

Lotus 1-2-3 (1983) – The Spreadsheet That Made IBM PCs a Business Standard

- Platform: IBM PC (MS-DOS)
- Why It Mattered:
    - Faster and more powerful than VisiCalc
    - Included charting and database features
    - quickly became the #1 spreadsheet program in business
    - Allowed finance professionals to run complex models without mainframes

Lotus 1-2-3 replaced VisiCalc as the standard finance tool and was a significant reason IBM PCs dominated corporate offices in the mid-1980s.

Microsoft Excel (1985) – The Start of a Monopoly

- Platform: Originally for Macintosh (1985), then IBM PC (1987)
- Why It Mattered:
    - First spreadsheet with a graphical user interface (GUI)
    - Could import and export financial data more easily than Lotus 1-2-3
    - Became dominant in the 1990s as part of Microsoft Office

At first, finance professionals stuck with Lotus 1-2-3 because it ran on MS-DOS. However, once Windows became popular, Excel overtook Lotus and became the default finance tool by the 1990s.

Early Accounting Software for PCs

As finance teams began to rely on PCs for spreadsheets, they also needed accounting software to handle transactions, invoices, and payroll.

Peachtree Accounting (1981) – First Business Accounting Software for PCs

- Platform: IBM PC (MS-DOS)
- Why It Mattered:
  - First full accounting suite for small and medium businesses
  - Allowed companies to manage payroll, accounts payable, and receivables
  - Helped companies move from manual bookkeeping to automated accounting

Peachtree was widely used in corporate finance departments and small businesses, helping transition from paper-based accounting to digital records.

QuickBooks (1983 - Developed by Intuit)

- Platform: IBM PC
- Why It Mattered:
  - Built for small business accounting
  - Simplified financial reporting, taxes, and invoicing
  - Making accounting accessible without needing a finance background

QuickBooks became one of the most popular financial software packages by the 1990s, eventually replacing many businesses' manual bookkeeping and ledger systems.

Great Plains Accounting (1984) – The First ERP for Finance

- Platform: MS-DOS, later Windows
- Why It Mattered:
    - First mid-sized business accounting ERP
    - Helped companies integrate accounting, payroll, and financial reporting
    - Acquired Microsoft in 2001 and became Microsoft Dynamics GP

Great Plains was an early competitor to SAP and Oracle in finance and introduced enterprise-level financial management on PCs.

Banking and Investment Software

As finance departments adopted PCs, banks and investment firms also moved away from mainframes for trading, financial modeling, and portfolio management.

Multiplan (1982) – Microsoft's First Finance Tool

- Platform: IBM PC, Apple II
- Why It Mattered:
    - Microsoft's first attempt at financial modeling software
    - Competed with Lotus 1-2-3 but was eventually replaced by Excel

Multiplan wasn't as popular as Lotus 1-2-3, but it paved the way for Excel's dominance in financial analysis.

Bloomberg Terminal (1982) – Financial Data for Traders

- Platform: Custom IBM-compatible system
- Why It Mattered:
  - Provided real-time financial market data
  - Used by investment firms, banks, and traders
  - Became a standard for stock market analysis and portfolio management

The Bloomberg Terminal revolutionized the financial industry by giving investors instant access to market data, replacing printed stock reports and slow mainframe queries.

Quicken (1983) – Personal Finance for Consumers

- Platform: IBM PC, Mac
- Why It Mattered:
  - First personal finance software for managing budgets and bank accounts
  - Allowed users to track expenses, income, and savings
  - Laid the foundation for online banking

By the late 1980s, Quicken was widely used by individuals and smaller finance teams to track business and personal expenses.

Early Enterprise Resource Planning (ERP) for Finance

As finance teams needed more automation and integration, early ERP systems helped businesses manage finances on a scale.

SAP R/2 (1979-1992) – The First Enterprise Finance System

- **Platform:** Originally mainframe-based, later adapted to client-server models
- **Why It Mattered:**
  - Allowed large companies to manage finance, HR, and logistics in one system
  - Transitioned from mainframes to PC-based enterprise computing
  - Became the foundation for modern financial ERPs

SAP R/2 was one of the last major finance software products to transition from mainframes to PC-based servers in the 1990s.

The 1980s Finance Software Revolution

Finance was one of the first industries to move to PCs and decentralized computing, mainly because of:

1. Spreadsheets (VisiCalc, Lotus 1-2-3, Excel) – Enabled real-time calculations.
2. Accounting Software (Peachtree, QuickBooks, Great Plains) – Allowed businesses to move away from paper-based bookkeeping.
3. Banking & Investment (Bloomberg Terminal, Multiplan) – Made financial market analysis more accessible.
4. ERP Systems (SAP R/2, early Oracle products) – Integrated enterprise-wide financial management.

By the early 1990s, PCs had completely replaced mainframes in finance departments, paving the way for modern fintech and cloud-based accounting

The 1990s marked a period of rapid transformation in finance technology as PCs completely overtook mainframes in financial departments. During this decade, we saw:

- The rise of ERP (Enterprise Resource Planning) systems to centralize financial operations.
- The beginning of cloud-based banking and trading.
- The first software war for PC dominance— Quattro Pro vs. Excel and Microsoft Office.

This decade set the stage for modern finance, where software, not hardware, determined efficiency, profitability, and market control.

The Spreadsheet Wars: Quattro Pro vs. Microsoft Excel

Before Microsoft Excel became the dominant finance tool, Lotus 1-2-3, Quattro Pro, and Excel were in a heated battle.

Lotus 1-2-3 Falls from Power

- Lotus 1-2-3 dominated the 1980s but struggled to transition to Windows.
- While it worked well in MS-DOS environments, it didn't fully embrace the graphical interface of Windows fast enough.
- By the early 1990s, businesses were moving to Windows-based finance software.

Quattro Pro: Borland's Challenger to Excel (1990)

- Developed by Borland, Quattro Pro was the first spreadsheet with a GUI, beating Excel to the punch.
- It introduced features like tabs (years before Excel added them).
- It was cheaper than Excel and had a strong user base.

But Quattro Pro made one mistake—it used reverse engineering to copy some Lotus 1-2-3 features. This led to lawsuits, and Quattro Pro lost momentum.

Microsoft Excel Overtakes the Market (1993-1995)

- Excel was part of Microsoft Office, which bundled Word, Excel, and PowerPoint.
- Integration with Windows 3.1 and Windows 95 gave Excel a considerable advantage.
- Microsoft aggressively marketed Office as a package, making it more cost-effective than buying separate software.
- By 1995, Excel had won. Lotus 1-2-3, and Quattro Pro was no longer a serious competitor.

Result: Microsoft crushed Lotus 1-2-3 and Quattro Pro by making Excel the default finance software in Windows.

The Rise of Microsoft Office and the Fall of Competitors

The 1990s saw the first significant software war on PCs. Before Microsoft Office dominated, businesses had many choices:

- WordPerfect vs. Microsoft Word
- Quattro Pro vs. Excel
- Lotus Notes vs. Outlook

Microsoft bundled its applications into Microsoft Office, forcing businesses to choose one package rather than mix software.

The Key Advantage of Microsoft Office

1. Lower Cost – Buying Excel, Word, and PowerPoint separately was expensive, but Microsoft bundled them cheaply than competitors.
2. Windows Integration – Since Microsoft made Office, it ran faster and smoother on Windows.
3. Corporate Licensing – Microsoft sold massive licenses to businesses, making it easier and cheaper to standardize Office across companies.
4. Elimination of Competitors – By the late 90s, Microsoft Office had destroyed:
   - WordPerfect in word processing.
   - Lotus 1-2-3 in spreadsheets.
   - Lotus Notes in email and productivity.

By 1997, Microsoft Office was the standard—and competitors like Lotus and Borland were fading.

This shift was driven by the need for companies to get on the same page when it came to sharing files. Back then, software like Microsoft Office, Lotus 1-2-3, and Quattro Pro weren't compatible with each other. File formats were proprietary, and trying to send a document between systems often led to failures or unreadable files. I clearly remember having to resend a file to another company after a user had to completely recreate it in Lotus 1-2-3, just so it would open on their end. These kinds of issues highlighted the need for standardization—and it's part of what pushed so many companies toward adopting Microsoft Office as the common platform.

Even Apple, at the time, struggled to integrate into this growing standard. Apple computers didn't come pre-installed with Microsoft Office, making them less practical for business use. That lack of compatibility combined with different protocols and sluggish network performance made Apple systems feel out of step with enterprise needs. While innovative in design, they simply didn't align with the direction most companies were heading, which only reinforced Microsoft's dominance in the workplace.

The Growth of Enterprise Financial Systems (ERP) in the 90s

While spreadsheets were great for individual finance teams, large corporations needed full accounting and financial planning systems. This led to the ERP boom.

SAP R/3 (1992) – The First Modern ERP

- SAP R/3 allowed businesses to manage finance, supply chain, and HR in one system.
- It was built for client-server environments, replacing old mainframe accounting systems.

- Major banks, manufacturers, and global businesses adopted SAP to streamline financial operations.

Oracle Financials (1990s) – The Banking Favorite

- Oracle was the leader in database technology, so banks trusted its financial solutions.
- Oracle ERP became the backbone of corporate banking and investment firms.

PeopleSoft (1987-1990s) – Financial Software for HR and Payroll

- Specialized in HR and payroll processing, later expanded into financial planning.
- Became a serious competitor to SAP until it was acquired by Oracle in 2005.

By the **late 1990s**, **SAP and Oracle dominated enterprise finance**—and they still do today.

The First Online Banking and Financial Trading Systems

As the internet grew, finance professionals needed real-time data access, leading to:

- Online banking and bill pay (first introduced in 1994).
- Web-based stock trading (E*TRADE launched in 1992).
- Online financial reporting tools that pulled data from ERP systems.

The Bloomberg Terminal Goes Online (1990s)

- Bloomberg, the gold standard for financial market data, transitioned from dedicated hardware to an online service.
- This allowed investment firms and traders to access market data from anywhere.

QuickBooks and Small Business Finance Moves to Windows (1998)

- QuickBooks launched a Windows-based version, replacing MS-DOS accounting software.
- By the late 90s, QuickBooks became the top small business accounting tool.

Banks Move to Digital

- Major banks began offering online access for customers to check balances and make transactions.
- By 1999, over 10 million people were banking online.

The End of Mainframes in Finance (1995-2000)

By the late 90s, PCs, ERPs, and online finance tools had entirely replaced mainframes in most finance departments.

- Banks still used mainframes for backend processing, but customer interactions moved to PC-based applications.
- Financial trading became fully digital, with traders using real-time stock market data on PCs.

- Corporate finance moved to SAP, Oracle, and PeopleSoft, eliminating the need for old batch-processing systems.

The Key Takeaways from the 90s:

1. Excel crushed all spreadsheet competitions (Lotus 1-2-3, Quattro Pro).
2. Microsoft Office dominated business computing, leaving WordPerfect and Lotus Notes behind.
3. SAP and Oracle became the kings of enterprise financial software.
4. Online banking and stock trading changed finance forever.
5. Mainframes were no longer needed for finance operations—PCs and servers replaced them.

The Legacy of the '90s: How It Shaped Modern Finance

Everything we see in modern fintech (PayPal, online investing, mobile banking) started in the 1990s.

- Microsoft Office is still dominant—Excel remains the most used finance tool worldwide.
- SAP and Oracle still control enterprise finance with cloud-based ERP solutions.
- Online banking exploded, leading to modern fintech platforms like PayPal and Venmo.

The software war of the 1990s was one of the most pivotal moments in business computing. It showed that whoever controls the software controls the business world.

1995-2000: Software Wars and Enterprise Transformation

The 1990s saw the rise of software that shaped the modern enterprise:

- Lotus 1-2-3 vs. Excel – Excel eventually won.
- WordPerfect vs. Microsoft Word – Word-dominated business documents.
- Novell NetWare vs. Windows NT – Microsoft took over networking.

As I worked across different companies, I saw the same trend everywhere—those who embraced PCs thrived, and those who resisted struggled.

My First PC and Pioneering the Way

My first real introduction to modern computing and the internet came when I bought a PC at Best Buy in 1996. It was running Windows 95 and had a 9.6K modem built in. I don't remember the exact make, but I believe it was an Aspire. When I set it up, I realized this was a completely different experience from anything before using the command line. Windows 95 was a drastic shift from Windows 3.1, and getting familiar with it took time. So, for a few months, I immersed myself in my new computer, exploring Windows 95, getting online with AOL, and figuring out how everything worked.

Building My First Website and Learning Web Development

As I got comfortable with Windows 95, I started experimenting with web development. Using Microsoft FrontPage, I launched a website called subfrequency.com, a platform for bands to sign up on and where I would manage their bookings. I marketed the site through AOL chat rooms and message boards, and to my surprise, a few bands signed up.

As the site gained traction, a brand-new online service called CD Now offered me a partnership opportunity. I regrettably turned it down and didn't realize how big that could be then. Eventually, I moved on from the project due to other opportunities, but this was my first real taste of the Internet's potential and how it was about to change business forever.

From Layoff to Lift-Off

In 1994, as things had really improved financially from my $10,000 raise and bonuses over a couple years, I started traveling again, something I have loved since childhood. My dad owned a motor home when I was young, and we camped all along the East Coast. That sense of freedom and exploration never left me. Now with a little extra in my pocket, I travel lightly in the past on my own (Family trips, Morocco, Bahamas) but now I can start seeing the world. Places I visited later in life: Germany, Austria, Dubai, Spain, Iceland, Australia, Guatemala, Costa Rica, the Caribbean, 70% of the states and during this chapter of my life, a trip to Mexico.

That trip was meant to be just that—fun and relaxing. But what happened there turned into something far more meaningful, something I carry with me to this day. If you know me, you know I've had some surreal encounters in my life. I sometimes joke that I've lived a little like Forrest Gump—crossing paths with interesting people and landing in situations you wouldn't believe unless you saw them. But this one? This one was divine.

While in Cancun, I decided to take a jet ski out with two of my friends. The sun was out, the water was clear, and everything felt perfect. One of my friends, riding in front of me, suddenly made a sharp turn, creating a huge wave that I couldn't avoid. I hit it hard—and the jet ski flipped.

From that moment on, everything went blank. I don't know how long I was under. It could've been a minute. Maybe two. I wasn't breathing, wasn't thinking—just silence.

But then, I opened my eyes. I was functioning normally. No choking. No panic. I saw the sun glowing circle at the surface—shining through twelve feet of water. I swam toward it, calm and fully conscious, and broke through the top like nothing had happened. Everyone back at the dock was staring wide-eyed, relieved. But I didn't think much of it. I climbed back on the jet ski and kept going.

Looking back, I realize I should have paused and acknowledged the moment. I am uncertain what saved me that day, but I remember clearly seeing the sun above, shining down through the water as if indicating that my journey was not yet over.

That moment—unexpected and unscripted was a wake-up call. We are not guaranteed our next breath, no matter how secure we feel. But God is always near, even in the middle of vacation, even in the middle of the ocean.

"The Lord will keep you from all harm—He will watch over your life." – Psalm 121:7

After this trip I was in for a shock. Before I stepped into my next major IT role, I went through a season of uncertainty that tested everything I thought I knew about career stability.

After working at a mortgage company where I gained invaluable hands-on experience with data communications, PC's, mainframes, terminal cabling, and early networking systems, I was laid off in June of 1994 during a massive corporate downsizing. It was a shock—but not a complete surprise. I had sensed a shift coming. The rise of personal computing, the automation of certain workflows, and the changing attitude in management were signs that technology was starting to streamline roles once seen as essential.

Until that point, I had been deeply involved in the planning and support of critical systems, keeping mainframes and terminals connected across the organization. I even guided a contractor who later ended up taking over my role, likely a more cost-effective move as the company looked to outsource what had become stable infrastructure. Despite all the overtime, late nights, and early-morning problem-solving I had contributed to in my early tech years, it felt like the foundation I helped build had been swept out from under me.

Then came the double blow: just one day after the layoff, I totaled my car in a crash. A drunk driver hit me as he ran a red light in DC. Now I have no job. No vehicle. No roadmap. It felt like everything was being stripped away but I believe it was God's way of clearing space for something new. What looked like a setback was the beginning of redirection.

So, I leaned into something I always carried with me: **music**.

*"For I know the plans I have for you... plans to give you hope and a future." – **Jeremiah 29:11***

I began playing music full-time with a couple of different bands, studio work, and I started my own sound and lighting company called Hollywood East. It wasn't consistent income, but it gave me something more valuable—a deeper education in hustle, responsibility, and logistics. I worked with some of the top promoters in D.C. at the time and some of the most popular A List groups, rappers and singers doing sound for their concerts and appearances. We had a contract with the best radio station in our city and things were going well.

I will never forget hauling massive speaker's up stairwells, setting up sound and lighting for large-scale events, and learning firsthand how event ideas move from a napkin to real life. I had to be on time. I had to be adaptable. And I had to deliver because there were no second chances when a live show was about to start.

That period taught me lessons no classroom or corporate training could ever teach how to manage pressure, how to see the whole system, and how to turn chaos into order, skills I would carry with me into enterprise IT.

Breaking into Enterprise IT: Becoming an IT Manager

In 1996, I got a job working as a PC and printer repair technician. I remember asking the interviewer for $28,000 a year trying to get close to what I made prior at the mortgage company, but I had been out of the IT world for two years. His response still echoes in my mind: *"Are you worth $28,000 a year?"* He offered me $23,500, and I accepted. It was all I had at the time.

I made the most of that opportunity. I worked hard, supporting some of the company's best clients and eventually surpassing the performance of more senior techs. I fixed one printer that had been sitting on the bench for 6 months which is what raised awareness to my talent. I traveled from site to site, learned how to oversee paperwork, collect customer signatures, and follow internal experiences that helped me understand the business side of technology.

Then one day, while working onsite for a client, someone took notice. They said I did great work—and asked if I'd be interested in a new opening at a beverage company. I interviewed for an IT Manager position at that beverage company, and I didn't pretend I hadn't been through a rough patch. I told them the truth: that I had been out of work, but I had used that time to sharpen my skills, learn Windows 95 inside and out, and explore the emerging world of the Internet.

That honesty, paired with everything I had learned on and off the stage, paid off.
I was hired. That moment reminded me that favor doesn't require perfection—just preparation. God opened that door when I had nothing left but faith and effort.

And that role became the next major steppingstone that launched me toward everything that followed.

*"A man's gift makes room for him and brings him before great men." – **Proverbs 18:16***

As IT Manager, I was responsible for guiding the company's evolving technology strategy, and one of my first major learning experiences came through Electronic Data Interchange (EDI), a system that proved to be a game-changer for our delivery operations. At the time, our beverage drivers used Norand handhelds to manage orders, track deliveries, and make log payments. Understanding how EDI worked wasn't just technical, it was operationally critical. It meant the difference between waiting hours to reconcile a route and doing it in near real time.

We only had one grocery store chain that supported full EDI transactions, but that one relationship showed me the power of automation. Their orders were often the largest, yet their delivery reconciliation was the fastest. I remember how those drivers consistently returned to the warehouse earlier than others. Why? Because there was no paper trail to manage—the entire transaction was electronic, from purchase order to delivery confirmation.

Back at the office, we used a cradle system for the Norand's, each one numbered and assigned to a specific driver. At the end of the route, the driver would insert the handheld into the appropriate cradle, which would upload the route data to our systems. Our cashiers who had recently transitioned from terminals to PCs—would then review the file. If there were still manual receipts from the route, they could adjust the data and finalize the reconciliation. But in the case of the EDI-supported customer, everything was already there—accurate and clean, requiring little to no manual intervention.

EDI eliminated the friction and guesswork. There was no missing receipt, no double entry, and no paper floating around. It was a closed-loop digital handshake between buyer and seller. Looking back, I see EDI as the early blueprint for what blockchain could become, a trusted, immutable, and near-instant framework for conducting business between organizations. Where EDI gave us speed and efficiency for one partner, blockchain could bring that same power to all vendors and buyers in the future, with inventory updates, payment settlements, and transactional integrity baked in from the start.

At the time, the company relied on a robust and reliable HP 3000 mini-frame system used throughout the business. The entire building was filled with terminal-based computers connected to this single machine. We had only one programmer who worked remotely and could dial in to make updates and program new features. I collaborated closely with her, ensuring our CFO and CEO had the support they needed for the company's IT operations.

Bringing Windows 95 to Enterprise: My First Major IT Initiative

With my experience using Windows 95 at home, I started envisioning how personal computers could improve our workplace. I approached the CFO and proposed a test project I wanted to bring in a PC and demonstrate how Windows could improve efficiency. To my surprise, he agreed, and we purchased a Compaq computer.

I configured the first machine, installing Windows 95 from floppy disks. This was my first time leading an IT transformation, and I then presented my case to all the head managers, CFO, and CEO, to purchase more PCs. It went incredibly well. The executive team saw the vision and the value, and soon, we started rolling out PCs across the office. Employees moved from terminals to Windows-based desktops, allowing greater flexibility and usability.

Installing Windows Server was another completely new experience for me. Back then, there were no real support systems, and books were scarce. I spent hours at the bookstore, reading everything I could find, learning about Windows 95, networking, and Windows Server administration.

Overcoming Resistance to Technology Adoption

One of the biggest challenges was getting employees to embrace the innovative technology. We see this even now with solutions like SharePoint. Many people resisted switching from terminals to PCs, either because they didn't want to learn something new or because they had been with the company for so long that they didn't see the need for change.

Some employees refused to take a computer, but management eventually made it mandatory as more departments adopted it. Even today, we see the same resistance when companies introduce new technology, AI, or automation—some employees embrace innovation, while others push back against change.

But learning is always beneficial. Gaining new skills opens doors to better opportunities, increases earning potential, and helps us stay relevant in a rapidly changing world. This should be a lesson for those who hate change, grasp it, and jump over it. Don't let things change around you; go with the change and be right on top of it.

My First Software Development Project: Customer Service Automation

We introduced a new Customer Service application as the company continued its technological transition. This was my first exposure to software development. A contractor was brought in to build the system, and I had to learn Windows Server and SQL Server to support it.

This project was complex and involved many moving parts, including boot order management and server cycling procedures. Within one year, I went from managing a handful of terminals to running four servers and supporting 14 PCs across the company.

Around this same time, in 1997, I also explored the idea of developing event management software. Inspired by the beverage software developer we worked with; I envisioned a comprehensive system that would significantly streamline the responsibilities of event planners. Although this project never fully took off, I meticulously outlined every feature and capability, believing deeply in its potential to transform the event management landscape. My passion for creating solutions and improving processes has always driven me forward, as it does for many others.

The Bible reminds us of our inherent creativity and purpose: "For we are God's handiwork, created in Christ Jesus to do good works, which God prepared in advance for us to do." (Ephesians 2:10 NIV)

Introducing Email to the Enterprise: The First Networking Challenges

One of the biggest game-changers was the introduction of email to the workplace. While email improved communication, it also came with unexpected challenges.

I installed Ethernet networking and deployed a small 8-port switch. At first, it worked well with just two or three computers, but as more machines were added, latency issues became a significant problem, especially with email.

My CEO was particularly frustrated. He once missed an important email because it was delayed, and I had to refresh the connection to retrieve it manually. After troubleshooting, I discovered the switch was the bottleneck—it simply couldn't manage the increased network traffic. We upgraded to a faster 16-port switch, which improved performance significantly.

Transitioning from The Beverage Company to IT Consulting

By 1998, my franchise had won the Best Franchise in the Nation Award, and I decided it was time for a change. After leaving my full-time role, I was a consultant for about two years, supporting the franchise's IT operations. Consulting was a great experience and a step up financially, and I had more time. More importantly, it allowed me to expand my user support and enterprise IT management skills.

Becoming a Field Tech: Lessons in Networking

Now that I have Windows OS and server management expertise, I started working as a traveling field technician for a friend's company. This was entirely new territory for me, I had done PC and printer repair before, but not on this scale, traveling to different businesses.

One of my first assignments was at a law firm that used a Star topology network. I wasn't familiar with Star Networks then and learned a harsh lesson the hard way. When I arrived, I disconnected one of the PCs, not realizing that every device relies on a central hub in a Star network. The entire network went down instantly.

The lawyer immediately started yelling at me, frustrated that their office had lost access to everything.

We quickly worked to restore the connection, but it was a valuable lesson—always understand the network architecture before making changes.

The Takeaway: A Decade of Learning and Growth

The late 1990s were a time of immense technological transformation. I had gone from buying my first PC at Best Buy to managing IT operations for a major corporation to consulting and fieldwork.

- The rise of Windows 95 revolutionized how businesses used technology.
- Networking and email adoption introduced new challenges but made communication more efficient.
- Enterprise IT shifted from terminals and mini frames to PCs and Windows Server environments.
- Learning on the job became essential as formal training resources were limited.

The Turning Point: From 1999 to the Next Era of IT

The uneventful passing of New Year's Eve 1999 wasn't just a relief—it marked the beginning of a new technological era. As the 2000s unfolded, IT was poised for rapid transformation. The rise of the Internet, cloud computing, and enterprise software systems would redefine businesses' operations. Companies that had spent the late '90s maintaining legacy systems were now forced to look ahead, preparing for a more connected and dynamic future.

One of the first significant shifts I witnessed firsthand was Novell's decline. Throughout the '90s, Novell's NetWare had been the dominant network operating system in many enterprises, providing file sharing, print services, and directory management. It was widely used, and I encountered it in company after company supporting IT environments that relied on it. But that dominance was short-lived.

2000 Microsoft launched Windows 2000, and with it came Active Directory—a significant change. Windows 2000 didn't just introduce a new operating system; it offered an integrated directory service that competed with Novell's core functionality. Almost overnight, companies began shifting away from Novell, realizing that Microsoft's ecosystem provided better integration with their growing IT infrastructures.

The decline of Novell was abrupt. Many businesses had been running mixed environments, using Novell for networking and Windows for applications. However, IT departments switched once Microsoft proved it could manage enterprise networks effectively without Novell. Over the next few years, I saw more and more organizations decommission their Novell servers, replacing them with Windows-based networks.

By the mid-2000s, Novell had faded into obscurity. The industry had moved on, embracing a new era of IT that prioritized integration, automation, and cloud computing. The shift away from Novell was just the beginning—technology was evolving faster than ever, and companies that failed to keep up were left behind.

The year 2000 wasn't just the end of Y2K fears; it was the start of an IT revolution. This began a much more significant transformation for those in the field.

The Beginning: Novell NetWare and the LAN Revolution

Novell was my next step. We used it to manage our computers and users on the network, which was probably the best way to manage organizational tech at that time. It was complex, and there was also hardware tools called sniffers. At some point, network traffic would collide, slowing things down. Using the sniffer helped us manage and troubleshoot errors.

Novell was one of the most prominent networking players during the 1980s and early 1990s. It pioneered Local Area Networks (LANs) before Microsoft even had a serious networking product.

- Founded in 1983, Novell introduced NetWare, one of the first server-based networking operating systems designed to link PCs together.
- NetWare provided file sharing, printer sharing, and directory services, which allowed businesses to centralize their IT infrastructure.
- Before Novell, networking PCs was impossible, you had to use direct connections, floppy disks, or mainframes.

How Novell NetWare Dominated the LAN Market

- IPX/SPX Protocol: Novell developed its proprietary networking protocol, IPX/SPX, faster than Microsoft's early TCP/IP implementations.
- Efficient Network Performance: NetWare was known for low overhead and high performance, making it a business favorite.
- Directory Services (NDS): Before Active Directory, Novell's NetWare Directory Services (NDS) was the gold standard for managing users, groups, and network access permissions.

135

By 1990, Novell owned 70% of the PC LAN market, while Microsoft was still figuring out Windows for Workgroups.

The Decline of Novell

1. The Rise of Windows NT Server (1993)
   o Microsoft released Windows NT Server, with built-in networking capabilities and native TCP/IP support.
   o Companies started moving to Windows because it integrated better with their desktops.
   o Microsoft bundled networking with its OS for free, while Novell charged for NetWare.
2. Novell's Strategic Mistakes
   o Novell refused to switch from IPX/SPX to TCP/IP early enough, making integrating with the growing Internet harder.
   o They acquired WordPerfect and Borland's Quattro Pro to compete with Microsoft Office, but these ventures failed.
   o Novell underestimated how quickly businesses would adopt Windows NT.
3. The Final Blow: Active Directory (2000)
   o When Microsoft released Active Directory in Windows 2000 Server, it completely replaced Novell Directory Services (NDS).
   o Businesses no longer needed NetWare because Windows Server could handle networking, security, and user management in one system.
4. Acquisition and Disappearance
   o By the mid-2000s, Novell was struggling. It was eventually acquired by The Attachmate Group in 2011 and later merged into Micro Focus.

- Today, Novell NetWare is completely obsolete—Windows Server, Linux, and cloud-based networking (Azure AD, AWS) dominate.

The Decline of Legacy Systems & The Y2K Turning Point

As the world counted to 2000, excitement and anxiety were in the air. The Y2K bug, a looming threat in the tech world, had been hyped as a potential disaster that could cripple financial systems, shut down power grids, and send industries into chaos. Businesses spent years preparing, testing, and patching their systems, hoping to avoid the worst.

At the time, I was working in IT for a significant beverage franchise company. Our infrastructure included an HP mini frame, which had been flagged as a system that *might* experience disruptions when the clock strikes midnight. My role was to help support the company's IT operations, and in the months leading up to Y2K, I collaborated with a programmer to evaluate the system and ensure it was ready for the transition. We ran simulations, checked for anomalies, and watched for any sign of failure—but nothing seemed out of place.

So, as the world braced for potential catastrophe on December 31, 1999, I found myself in a very different setting—on stage, playing bass at a club in Washington, D.C. While others in IT were glued to their screens, waiting for the first signs of failure, I was keeping an eye on my pager, expecting a call if anything went wrong. Midnight came and went. The music played on. My pager stayed silent.

This was my original stack—built for sound before I built for systems.

For us, Y2K turned out to be disappointing. No crashes, no disruptions, business as usual. But that was not the case for everyone. Some companies weren't lucky, and the industry learned a valuable lesson about the risks of legacy systems and the importance of proactive IT management.

Looking back, Y2K was a defining moment in IT history—one that shaped how we think about risk, preparedness, and the evolution of technology. For me, it was one of those rare moments when I could be both an IT professional and a musician, living in two different worlds on the night that was supposed to change everything.

The Y2K bug, or the Millennium Bug, was a principal software flaw that threatened computer systems worldwide as 2000 approached. Many legacy systems, especially those built on mainframes, stored years using only two digits (e.g., "99" for 1999). This meant that when 2000 (00) arrived, these systems might interpret it as 1900, leading to data corruption, calculation errors, and system failures.

Why Did the Y2K Bug Exist?

- Early computers had limited storage – Memory was expensive in the 1960s-1980s, so engineers shortened dates to save space.
- Two-digit years were a programming shortcut – No one expected these systems to last into the 2000s.
- Banking, finance, and government systems relied on old mainframe code—many were running COBOL, FORTRAN, and assembly language that had been unchanged for decades.

The Risks and Global Panic Leading Up to Y2K

By the mid-1990s, experts realized that thousands of mission-critical systems had this flaw. There were fears that:

- Banks and stock markets would fail due to date miscalculations.
- Planes might be grounded if navigation and scheduling systems fail.
- Power grids could shut down if utility companies' software malfunctioned.
- Government systems (Social Security, Medicare, IRS) could stop working.
- Nuclear reactors and military defense systems could misinterpret data, leading to security risks.

As a result, businesses and governments spent billions fixing the issue, launching massive Y2K compliance projects.

Actual Y2K Incidents: What Happened?

Despite the global panic, most critical systems were fixed in time. However, some smaller-scale Y2K failures did occur, including:

Banking & Finance Issues

- The U.S. Federal Reserve reported 60 minor Y2K-related failures but no central banking collapses.
- Bank of New York had a $700 million overdraft due to a Y2K glitch in its interest calculation system.
- U.K. credit card systems overcharged some customers due to errors in expiration date validation.

Airline & Transportation Failures

- In Norway, some train ticket machines stopped working, refusing to issue tickets.
- In Australia, bus ticketing systems failed, causing temporary free rides.
- The U.S. Naval Observatory clock miscalculated the date, briefly displaying "19100".

Power & Utility Issues

- In Japan, a nuclear power plant alarm went off at midnight due to a Y2K-related glitch, but no real issue occurred.
- In the U.S., a few electric grids had minor malfunctions, but power continued uninterrupted.

Government & Military Glitches

- In South Korea, some military satellites failed temporarily due to Y2K-related software bugs.
- U.S. spy satellites briefly sent incorrect data but were quickly corrected.

Medical & Public Safety Issues

- Some hospital equipment displayed incorrect data but did not fail.

- In the U.K., a pregnant woman was mistakenly recorded as being 100 years old in a hospital database.

No major catastrophic failures occurred, but the risk was real, and billions were spent fixing vulnerabilities.

How Much Did the World Spend on Y2K Fixes?

Governments and businesses spent an estimated $300–600 billion globally to prepare for Y2K.

- United States: $100 billion spent, with the federal government investing over $8 billion to ensure compliance.
- United Kingdom: $50 billion spent by businesses and government.
- Japan: $100 billion+ spent on fixing financial systems and utilities.

Most of the costs went towards:

- Rewriting old COBOL and FORTRAN code in legacy banking and government systems.
- Testing and replacing embedded systems in transportation, healthcare, and utilities.
- Performing audits and compliance checks to ensure companies met Y2K readiness standards.

Was Y2K Overblown, or Did We Prevent a Crisis?

The fact that no significant disasters happened led some to believe Y2K was a hoax or an overhyped crisis. However, many IT professionals argue that we avoid catastrophes because of massive preparation efforts.

Without the fixes:

- Many banking systems would have failed, causing financial crashes.
- Power plants could have experienced shutdowns due to date miscalculations.
- Stock markets could have miscalculated trading dates, leading to significant financial losses.

A Lesson in IT Preparedness

The Y2K scare showed:

1. Legacy software needs proactive maintenance – Many companies waited too long to modernize systems.
2. Governments & businesses need coordinated IT strategies – Large-scale IT risks require global cooperation.
3. The cost of prevention is worth it – Fixing software before failure is cheaper than dealing with a crisis.

The Long-Term Impact of Y2K on IT & Finance

Y2K led to significant changes in software development and IT risk management:

1. More frequent software audits – Businesses now regularly check for software expiration issues.
2. Better documentation of legacy code – Companies realized the risk of relying on old, undocumented mainframe systems.
3. The rise of ERP systems – Many businesses switched to SAP, Oracle, and Microsoft to modernize their financial software.

4. Increased focus on compliance – Governments introduced new IT security laws to prevent future crises.

What If Y2K Happened Today?

If a similar date-based flaw existed in modern cloud-based and AI-driven systems, it could cause:

- Banking failures due to algorithmic miscalculations.
- Stock market glitches causing AI trading crashes.
- Utility shutdowns from smart grid miscalculations.
- Healthcare system errors due to incorrect medical records.

However, today's software development standards include better testing, automated updates, and more robust compliance measures to prevent similar large-scale issues.

Looking Back on Y2K

- The Y2K bug was real, and without preparation, it could have caused financial and infrastructure chaos.
- The lack of significant failures was due to global coordination and massive IT efforts.
- Y2K taught businesses the importance of maintaining and modernizing software systems.

The real legacy of Y2K was that it forced the world to take IT risk management seriously, shaping the cybersecurity and IT compliance policies we use today.

By the time we entered the 2000s, the rules had changed. IT wasn't just a support function—it was the heartbeat of business. For me, what started with a floppy disk and a command line became a lifelong passion for bridging technology with purpose.

But the most significant changes were yet to come!

Chapter 6:

Defining the 2000s & 2010s: The Most Transformative Era in Tech

This next step in my journey marked a personal and professional transformation. The early 2000s ushered in a new era for technology, the rise of the internet, the dawn of web development, and the reshaping of enterprise IT systems. Everything was changing fast, and I was ready to change with it.

Before fully immersing myself in technology, I had spent years traveling around the city repairing computers and small office networks. But my heart was still split. I was also pursuing a career in music. For nearly two decades, I played bass professionally, poured my creativity into songwriting, and lived the life of a performing musician. That chapter ended on New Year's Eve, 2000, when I officially stepped off the stage to begin a new journey in tech.

The life of a musician is often misunderstood. It requires discipline, sacrifice, and relentless effort. I spent countless nights rehearsing, performing, and recording—often leaving little time for sleep, rest, or recovery. That schedule clashed heavily with IT work, especially in an era when systems maintenance happened late at night or over the weekend. I was constantly bouncing between two demanding worlds: one of structured systems, the other of raw improvisation.

And yet, I wouldn't trade those years for anything. Being a musician taught me more about technology than most people would expect. In those days, keyboards were connected via MIDI and getting them to sync was just like configuring a small network. Each device had its own menu system, its own protocols—and you had to know how to get everything by talking to each other in time. It was orchestration in its truest form.

Recording was another challenge altogether. We worked with magnetic tape, and editing required precision and creativity. Splicing a tape to remove a section—or build a new arrangement—wasn't just technical, it was intuitive. You had to feel the rhythm, cut on the beat, and stitch it back together so no one could hear the seam. It was both science and soul.

I often say this: never underestimate musicians. Especially the ones without formal training. Creativity, adaptability, and persistence are in their DNA. Musicians take risks. They build something from nothing. And most of what I have met know how to listen, a skill far too rare in leadership and technology today.

I do not share this to boast, I share it in reflection. I taught myself bass guitar. I played by ear. And by God's grace, I had the opportunity to play on tracks for some incredible artists, including Michael Jackson and Groove Theory. That experience taught me something I've carried with me ever since: your value isn't always about where you are, it's about what you carry. Sometimes, you're just waiting to be placed where you truly fit.

And let me be clear, this was not about a college degree or a stack of certifications. It was about focus, repetition, and belief. I didn't follow the traditional route, and I've learned that you don't have to. If you want something bad enough, you can learn it. You can build it. You can live it without conforming to the path others expect you to take. That principle has guided me in music, in technology, and now in how I lead.

Music gave me rhythm, but more than that it gave me structure. It taught me how to listen, how to adjust on the fly, how to lead without taking the spotlight, and how to coordinate chaos into harmony. And there is no better example of that than events.

From a youthful age, I've been part of event execution—from sitting at my dad's baseball games selling food to spectators, to later running sound and lights, booking bands, managing logistics, and loading gear at 2 a.m. I've done events from both sides of the stage—and behind the booth. Musicians improvise the impossible, and when you're the one coordinating gear, performers, timing, and vendors, you learn fast that a spreadsheet won't cut it.

Many of the events I helped plan were not born in boardrooms, they started on the back of napkins in loud rooms, with nothing but an idea and there are few people willing to make it happen. It was not glamorous. But it was real. And those experiences taught me more about execution, timing, and resource coordination than any system I've seen in the enterprise world. I understood what event management was because I experienced it, and I tried to create it in 1997, but it was always on my mind.

So, a couple of years ago I built **Tourbook**—a next-generation, AI-powered event and coordination platform. It's a modular component of **Tourque**, our AI Operating System, but it stands on its own as a complete event intelligence system—designed by someone who's lived every chaotic second of event management from the trenches.

Tourbook isn't about managing speakers and coordination. It's about capturing what really happens: energy, dependencies, the people, the vendors, the changes, the follow-up. It understands that events aren't checkboxes. They require rhythm, accountability, and visibility.

For years, I gave everything to music—hours of practice, late-night rehearsals, and the pressure of live performance. Then one day I asked myself:
What if I applied that same level of focus and discipline to technology?
Where could it take me?

God gives us gifts—and He honors them when we use them with intention. Music trained my mind; technology gave me my mission.

*"Whatever your hand finds to do, do it with all you might." – **Ecclesiastes 9:10***

That question launched my full transition into IT. From 2000 to 2014, I committed myself entirely to learning, building, and evolving in the tech space. I rode the wave of change as the industry shifted from local networks to cloud computing, from static websites to dynamic applications. It was one of the most exciting periods in modern computing and I was all in.

What I saw and learned during those years shaped not only my career, but how I understand systems, people, and the world we build with technology.

🎶 What Music Taught Me About Technology

"Every gig, every rehearsal, every recording session gave me something that later showed up in tech."

🎸 From Music	💻 To Technology
Practicing scales & songs daily	Repeating system builds & script testing
Playing live without a set list	Jumping into new environments with minimal prep
Syncing MIDI instruments	Configuring and troubleshooting networks
Splicing magnetic tape on beat	Debugging, refactoring code, and system patching
Jamming with unpredictable bandmates	Collaborating with complex teams and personalities
Improvising under pressure	Adapting during high-stakes projects or outages
Listening to the rhythm	Understanding timing, flow, and systems behavior
Creating from nothing	Architecting solutions from scratch

Creative people aren't just visionaries, they're survivors

Musicians know how to repeat, refine, and risk. That mindset didn't leave me when I put down the bass it evolved with me as I stepped into data centers, boardrooms, and code.

The Web Development Phase: Mastering Flash and Building Websites

One of the areas I continued to focus on after my time in the beverage industry was website development. I loved the creativity and technical challenges of building websites, and I quickly realized this was an area where I could both innovate and generate income.

Flash was one of the most powerful tools available for web development during this time. Initially designed as an animation platform, Flash became an essential interactive web design tool. I took the time to master Flash, learning how to animate, build timelines, and create immersive web experiences. The ability to make things slide, fade, and move dynamically made Flash an exciting technology.

I still have all my old Flash sites saved on my external drive, and sometimes, I look back at them as a reminder of how far technology has come. I had the opportunity to build websites for notable figures like:

- Marcus Johnson, a famous jazz keyboardist,
- Biz Markie, where I was the webmaster for his site for many years,
- Countless other websites for work, friends, and family.

Flash was challenging to learn because it was based on a timeline-based structure, similar to video editing. Since Flash was all about movement, you had to plan animations, transitions, and interactions carefully. Despite its steep learning curve, I became proficient in it, and it remained a core skill for many years—until Apple's decision to phase it out changed everything.

Landing a Job at a Government Contracting Firm: Expanding into Enterprise IT

After being a traveling PC repair technician and I continued to explore web development and online business, I was also looking for opportunities to grow in the enterprise IT space. In 2001, through a friend, I landed a job at a government contracting company, a significant technology and leading tech company at the time.

This was a turning point in my career. The government contracting company gave me exposure to enterprise IT systems, where I learned:

- SQL Server – Database management and enterprise data storage.
- Windows Server – The backbone of corporate IT environments.
- Documentum – Enterprise content management system for document storage.
- FileNet – A workflow and document management platform widely used in government and large corporations.

Working at a government contracting company piqued my interest in SQL Server, and I became highly proficient in supporting various enterprise applications and databases. Windows Server and Active Directory also became core strengths, allowing me to build a strong enterprise IT administration and infrastructure management foundation.

Understanding the Direction of Technology

The early 2000s were crucial in tech. The internet was exploding, businesses were transitioning to online platforms, and IT was rapidly evolving.

At the government contracting company, I began seeing the larger picture of how technology shaped the future. Everything was moving toward digital transformation, data centralization, and workflow automation. What started as a simple curiosity about web development has now led me to the heart of enterprise IT.

This period set the stage for my next technological phase, where I would go beyond managing systems and start architecting solutions for businesses, government agencies, and organizations navigating the new digital era.

The 2000s and 2010s were decades of unparalleled technological change. They marked the shift from traditional computing to cloud-based services, the decline of legacy hardware, mobile computing, and explosive internet speed and connectivity growth. Companies that had dominated computing in the '80s and '90s faced new challenges, and a new set of industry leaders emerged.

Where Was Technology in the 2000s?

The early 2000s were an awkward transition period when businesses still used legacy systems, but the internet was rapidly evolving. Computing power increased, broadband became the standard, and Wi-Fi replaced wired networks.

Faster Computer Chips & The Rise of Multi-Core Processing

The early 2000s saw a significant increase in processing power. Companies shifted from single-core CPUs to multi-core architectures, making computers exponentially faster.

- Intel Pentium 4 (2000) → Single-core CPU with clock speeds over 3 GHz.
- AMD Athlon 64 (2003) → First 64-bit consumer processor, leading to improved performance.
- Intel Core Duo (2006) → The first mainstream dual-core processor, making multitasking much smoother.
- Intel Core i7 (2008) → Ushered in the era of hyper-threading, allowing one physical CPU to function as two logical processors.

Internet Speeds Take Off: The Shift from Dial-Up to Broadband

The 2000s marked the death of dial-up internet and the rise of broadband connections. Businesses ditched T1 lines for fiber and high-speed DSL, dramatically changing their operations.

- Dial-up (2000s) → 56 kbps using a telephone line.
- DSL (Early 2000s) → 256 kbps - 3 Mbps (game-changer for businesses and consumers).
- Cable Internet (2005-2010) → 5-50 Mbps speeds, powered by ISPs like Comcast and Time Warner.
- Fiber Optic Internet (Late 2000s) → Google Fiber, Verizon Fios, and AT&T U-Verse brought speeds of 100 Mbps+.

- Wi-Fi Expansion (2000s) → The introduction of 802.11a/b/g enabled wireless internet in homes and offices.

The Fall of Dial-Up & The Modem

By the late 2000s, the modem was disappearing. Once a crucial piece of tech for connecting to the internet, it became obsolete as broadband took over. The 'handshake' sound of dial-up modems became a relic of the past.

- AOL and EarthLink—once the kings of dial-up—collapsed as broadband took over.
- Routers replaced modems as wireless networks became the new standard.

The rise of cloud computing and streaming services has made the internet necessary.

The PC War: Dell, Compaq, and HP

The early 2000s were a golden age for desktop computers, but the competition was fierce.

- Dell (2000s) → Dominated corporate computing by offering cheap, customizable PCs.
- Compaq (2002) → Acquired by HP, marking the beginning of the end for Compaq as a brand.
- HP (2000s-2010s) → Became the largest PC manufacturer after absorbing Compaq.
- Apple (2000s) → Began to make a comeback with the success of the iMac (1998) and MacBook (2006).

The rise in laptops and mobile computing led to the decline of desktops in the late 2000s.

Where Were Bill Gates & Steve Jobs in the 2000s?

Two of the biggest names in computing—Bill Gates and Steve Jobs—entered very different chapters of their careers during this era.

Bill Gates: The End of His Microsoft Era (2000s)

In 2000, Gates stepped down as CEO of Microsoft, handing the role to Steve Ballmer.

By 2006, he announced his retirement from full-time Microsoft work to focus on philanthropy.

In 2008, he officially left his day-to-day role at Microsoft and shifted focus to The Bill & Melinda Gates Foundation.

However, Gates' early 2000s contributions shaped modern computing:

- Windows XP (2001) → One of the most successful operating systems ever.
- Microsoft SharePoint (2001, 2003 & 2007) → Transformed business collaboration and document management.
- Microsoft Office 2003/2007 → Introduced the ribbon UI and XML-based file formats (.docx, .xlsx).

Steve Jobs: Apple's Comeback (1997-2011)

After returning to Apple in 1997, Jobs completely reshaped the company.

- iMac (1998), Mac OS X (2001), and iPod (2001) brought Apple back into the spotlight.

- The iPhone (2007) and App Store (2008) revolutionized the mobile industry.

Unfortunately, Jobs's health declined, and he passed away in 2011, but by then, Apple had become the most valuable tech company in the world.

What Happened to IBM?

Once the dominant computing force, IBM faced serious challenges in the 2000s.

- 2002 → IBM sold its hard drive business to Hitachi.
- 2005 → IBM sold its PC division to Lenovo, exciting the consumer PC market.
- 2000s-2010s → IBM pivoted to enterprise services, AI, and cloud computing (Watson, Red Hat acquisition).
- 2020s → IBM became a cloud computing and AI powerhouse but no longer a consumer tech company.

The Rise of Identity Management & Cybersecurity

As more businesses went digital, identity management became critical.

- Microsoft Active Directory (2000s-2010s) became the standard for managing enterprise users.
- Okta & SSO (Single Sign-On) Solutions (2010s) changed how users accessed applications.
- 2FA (Two-Factor Authentication) and MFA (Multi-Factor Authentication) became standard security measures.

- ACTi Identity (early 2000s) pioneered smart card–based authentication, especially in government and defense, laying the groundwork for secure identity management and PKI

With the rise of cloud computing, companies needed stronger authentication and role-based access controls to protect data.

Software Wars: Microsoft vs. Adobe vs. Apple

In the mid-2000s, new software battles emerged.

- Adobe Flash vs. Apple (2007-2010) → Apple banned Flash from iPhones and iPads, killing Flash.
- Microsoft Office vs. Google Docs (2006-Present) → Google disrupted Microsoft's dominance.
- Adobe vs. Microsoft → Adobe challenged Office with PDF technology and subscription-based software (Adobe Creative Cloud).

The introduction of SharePoint MOSS 2007 became a massive shift in enterprise collaboration. SharePoint allowed companies to digitize their document management, creating a foundation for Microsoft Teams, OneDrive, and Office 365 in the 2010s.

The Explosion of E-Commerce & Online Ordering

Amazon changed how people shopped.

- Amazon Prime (2005) → Launched two-day shipping, setting a new standard.
- eBay, PayPal, and digital payments exploded.

- E-commerce surpassed brick-and-mortar retail by the late 2010s.

The Mobile & Cloud Era (2010s)
By 2010, everything went mobile, and cloud based.
- Smartphones replaced PCs as the primary computing device.
- Social media (Facebook, Twitter, Instagram) exploded.
- Streaming services (Netflix, YouTube, Spotify) dominated entertainment.
- Cloud computing (AWS, Azure, Google Cloud) replaced local servers.

The internet became fully mobile, and companies that didn't adapt were left behind.

The Rise and Fall of the Microsoft Windows Phone

I remember working at a top technology company when they introduced the Windows Phone, which we could use for work. I loved the phone, which had many advanced features over other phones, but it entered the race too late.

When Microsoft launched the Windows Phone platform in 2010, it entered a mobile battlefield dominated by Apple's iOS and Google's Android. With its sleek "Live Tile" interface, deep integration with Microsoft Office, and emphasis on enterprise connectivity, the Windows Phone promised a fresh alternative to the traditional app grid. For users already immersed in the Microsoft ecosystem—particularly businesses—the phone appeared to offer a seamless mobile extension of the desktop experience.

At its core, Windows Phone had a few strong differentiators:

- Live Tiles provided real-time updates briefly, giving the home screen a dynamic, personalized feel.
- It offered native integration with Outlook, Office, OneDrive, and SharePoint, making it attractive to enterprise users.
- Nokia, once a mobile titan, partnered with Microsoft to produce high-quality Lumia hardware that ran the Windows OS.

Despite these advantages, the platform never gained meaningful market share. By 2013, Windows Phone held just 3.6% of the global smartphone market, which would continue to decline. The reasons for its failure were multi-faceted:

Why the Windows Phone Failed

- **App Ecosystem Deficiency**
  Perhaps the most fatal flaw was the lack of apps. Developers prioritized iOS and Android, where the user base was more extensive, and monetization opportunities improved. Even significant apps like Instagram, Snapchat, and YouTube were missing, outdated, or released late on Windows Phone. For consumers, this gap in functionality was a deal-breaker.
- **Too Late to the Game**
  Microsoft arrived several years after Apple and Google established mobile dominance. When the Windows Phone gained traction, the iOS and Android ecosystems matured and became deeply entrenched.

- **Confusing Brand Strategy**
  Microsoft's branding and platform strategies were inconsistent. It rapidly shifted from Windows Phone 7 to 8 to 10, disrupting compatibility and confusing developers and consumers each time. Many early adopters were left with unsupported devices just a year or two after purchase.
- **Poor Developer Support**
  Microsoft tried to lure developers with tools like Xamarin and universal app frameworks, but many found the learning curve steep or the return on investment too low. Without critical mass or strong incentives, developers moved on.
- **Nokia Acquisition Misfire**
  In 2013, Microsoft acquired Nokia's handset division in a bold attempt to control hardware and software. While Lumia phones were well-built and innovative, the acquisition cost billions of dollars and failed to change market dynamics. Microsoft eventually wrote off the acquisition, laying off thousands of employees.
- **Lack of Consumer Demand**
  Consumers didn't ask for Windows Phones. While the devices were capable and often offered excellent cameras (especially in the Lumia line), they didn't offer anything compelling enough to sway users from Apple or Android.

The Legacy of the Windows Phone

Despite its commercial failure, Windows Phone left a mark on the industry:

- Its Live Tile interface influenced future Windows desktop and tablet designs.

- It pioneered the idea of universal apps, which would later shape Windows 10.
- It reminded Microsoft—and the tech world—that innovation without widespread ecosystem support will struggle to survive.

In 2017, Microsoft officially announced the end of the active development for the Windows Phone. By 2020, the platform was dead, with app store support removed and updates discontinued. The failure to break into mobile was a significant turning point for a company that once dominated the PC world.

My Thoughts on Windows Phones:

I remember being at a top tech company when the Windows Phone launched. As employees, we had the choice to use it for work—and I did. To this day, I'll say it: I loved that phone. It was ahead of its time in so many ways. One feature that stood out to me was voice-to-text messaging. 2012, I could answer text messages just by speaking and it worked. Seamlessly. Fast forward to today, and I still haven't seen that level of voice consistency replicated across platforms like iOS or Android.

The Live Tiles felt fresh, the integration with Outlook and Office was smooth, and the overall experience was clean and intuitive. It felt like the future.

But even the best tech can stumble when it enters the race too late. The Windows Phone had potential, but it was released into a market already dominated by iPhone and Android—two ecosystems with strong developer support and massive user bases. As much as I appreciated the innovation, the reality of the market was less forgiving.

Windows Phone wasn't a bad product; it was a good idea launched at the wrong time with the wrong strategy. In a world where ecosystems matter as much as innovation, Microsoft's mobile ambitions were outpaced by faster, more flexible competitors.

The Most Disruptive Decades Ever

The 2000s and 2010s were the most transformative eras in technology. The old tech giants (IBM, Compaq, AOL) declined, while Apple, Google, Amazon, and cloud computing took over. Faster speeds, mobile computing, and the rise of identity management have changed how we interact with technology.

This era set the stage for modern AI, automation, and the next phase of digital transformation—the world we live in today.

The Dot-Com Boom & Bust

A Shift in Technology and Jobs

The Dot-Com Boom, which started in the mid-to-late 1990s, was fueled by an explosion of internet-based businesses, venture capital investments, and e-commerce. Companies were rushing to establish their online presence, betting on the web's potential to revolutionize industries. This era was of tech startups, inflated stock prices, and rapid innovation in digital communication, software, and web-based services.

By the late 1990s, businesses had embraced the Internet to cut costs, automate processes, and transition to digital workflows. Previously highly specialized, labor-intensive, and profitable industries—such as printing, publishing, and design—began experiencing a significant shift. Once highly paid for their expertise in print production, graphic designers saw the emergence of software like Adobe Photoshop, Illustrator, and InDesign, which enabled companies to produce high-quality designs in-house without outsourcing expensive printing services.

The Dot-Com Bust (2000-2002): The Collapse of the Tech Bubble

By March 2000, the Dot-Com Bubble burst, causing the NASDAQ stock market to plummet from 5,048 points to a low of 1,139 by October 2002. This 80% decline wiped out billions of dollars in investments, led to massive layoffs, and caused many once-promising tech companies to shut down overnight. Although we are led to believe that AI will not take jobs, I think we will see some similar things happen with AI.

During this period, industries that relied on printing, advertising, and traditional media faced steep revenue declines as digital tools became more powerful and widespread. Companies no longer needed to spend hundreds of thousands of dollars on printing services when they could generate designs digitally and distribute them online. The result? Jobs that once paid six figures were now worth a fraction of what they once were.

My friend's ex-boyfriend perfectly illustrates this transition. He made $175K a year in the printing industry, and it was once a solid, lucrative career. However, when high-end digital printers, online ordering systems, and Adobe software became accessible to the masses, the need for traditional, outsourced print jobs disappeared overnight. His income dropped from $175K to $40K in just a few months as businesses slashed budgets, canceled contracts, and moved their design and print production in-house.

The Dot-Com Bust didn't just impact tech startups—it had ripple effects across multiple industries, especially those tied to traditional advertising, marketing, and print production. Many graphic designers, print shop owners, and publishing professionals saw their careers take significant hits, forcing them to adapt to the digital revolution or risk becoming obsolete.

The Shift from Print to Digital

As companies moved their operations online, the demand for traditional print media dropped significantly. Here are some of the most significant shifts during this era:

1. High-end printing became a Commodity – Companies no longer needed specialized and large-scale offset printing when digital presses and home-office printers could produce quality prints at a fraction of the cost.
2. Adobe & Digital Design Took Over—Software like Adobe Photoshop, Illustrator, and InDesign allowed businesses to eliminate the need for print shops and in-house designers.
3. Web & Online Marketing Replaced Print Advertising – Companies shifted from physical marketing materials (brochures, catalogs, direct mail) to websites, email marketing, and social media.
4. E-Commerce Became the Norm – Businesses moved from brick-and-mortar operations to online platforms, reducing the need for physical signage, banners, and traditional advertising materials.

The Aftermath: Adapt or Die

By 2003, the tech industry began to recover, but the traditional printing industry never fully rebounded. Many highly paid professionals had to improve their skills in digital design, pivoted to new industries, or take significant pay cuts. The digital transformation era had officially begun, forcing professionals across industries to adapt to an online-first world.

This moment in history was a wake-up call—if technology could disrupt one of the most stable, high-paying industries like printing, then no industry was safe from digital transformation.

I see the same pattern forming again—but this time, the shift is even more intense. As AI advances rapidly across technologies and industries, we're entering a new era of computing that doesn't just reward innovation demands reinvention. This is no longer about adopting the next shiny tool. AI is a new operating model—a smarter, faster way to do business. Those who adapt will accelerate. Those who hesitate risk being caught in a tailspin they won't recover from.

Don't get left behind.

The Rise of Workflow and Automation in the 2000s

The 2000s marked a revolutionary shift in IT, as companies began embracing workflow automation to streamline processes, reduce manual tasks, and enhance collaboration across departments. At the government contracting company, I worked as a FileNet/SQL Admin and I witnessed workflow automation in action for the first time, changing how I viewed technology forever.

We implemented a new way of doing business in those days at the government contracting company. The team implemented a workflow system that utilized multiple servers, all communicating with each other through custom code and off-the-shelf applications. This system allowed documents to be routed through automated approval chains, ensuring that the right stakeholders reviewed and approved critical business processes without unnecessary delays.

It was eye-opening to see servers orchestrating tasks independently, moving data between systems without human intervention. This was the earliest form of modern automation, long before tools like Microsoft Power Automate or Zapier made process automation accessible to everyone.

How Workflow Automation Changed Business Operations

Before automation, many business processes relied on paper-based approvals, manual data entry, and email-based workflows requiring employees to track approvals and follow-ups manually. This led to inefficiencies, bottlenecks, and errors in critical processes.

The workflow system we implemented transformed these operations by:

- Eliminating manual approvals—Documents and requests move through a predefined digital approval chain instead of waiting for an email response, ensuring faster turnaround times.
- Integrating multiple systems – The servers were programmed to exchange data between various applications, reducing the need for employees to

manually transfer information from one system to another.

- Enhancing compliance and tracking – Every action within the workflow was logged, ensuring a clear audit trail for accountability and regulatory compliance.

This experience profoundly impacted me, making me incredibly curious about the potential of automation and how technology could transform business processes.

The Foundation for Modern Automation Tools

At the time, workflow automation required custom development, and it was used by large enterprises with the resources to build these complex systems. But as technology evolved, automation tools became more accessible, paving the way for solutions like:

- Power Automate is Microsoft's low-code automation platform, which allows users to create workflows across multiple applications with minimal coding.
- SharePoint Workflow & Nintex – Early no-code workflow tools enabled businesses to build approval and document management systems.
- Robotic Process Automation (RPA) – Tools like UiPath and Automation Anywhere emerged to mimic human interactions with software applications, further pushing the boundaries of what could be automated.

Looking back, the workflow system I helped implement at the government contracting company was an early version of what is now a multi-billion-dollar industry. It inspired my passion for automation, which would become a significant focus throughout my career, helping businesses improve efficiency, reduce errors, and accelerate decision-making through automated systems.

These tools were expensive. Some of the tools were:

Microsoft BizTalk Server (2000)

- One of Microsoft's early entries into enterprise application integration (EAI) and business process automation.
- Used for automating business processes and integrating disparate systems using XML and SOAP.
- Connected legacy systems, databases, and external trading partners.

IBM WebSphere Process Server (2005)

- Built on WebSphere Application Server, it provided workflow automation, service orchestration, and business process management (BPM).
- Used BPEL (Business Process Execution Language) to define workflows.
- Later, it became part of IBM Business Process Manager (BPM).

FileNet (Acquired by IBM in 2006)

- Initially developed workflow automation and enterprise content management (ECM) solutions.
- Focused on document-driven workflows and case management.
- A leader in the BPM market before being absorbed into IBM's portfolio.

TIBCO, I Process Suite

- TIBCO was a major player in middleware and integration.
- process Suite allowed for workflow automation and business process orchestration.
- Known for real-time event processing and high-performance workflow automation.

K2 (Founded in 2000)

- Developed workflow and BPM solutions using Microsoft technologies.
- Created tools for SharePoint-based workflow automation.
- Later, it evolved into a low-code process automation platform.

Ultimus (Founded in 1994, strong presence in the 2000s)

- A major BPM and workflow automation company.
- Known for business rules-driven workflows.
- Provided visual process modeling for workflow automation.

Staff ware (Acquired by TIBCO in 2004)

- A UK-based workflow automation company.
- One of the pioneers in BPM and enterprise workflow systems.
- Integrated into TIBCO's BPM offerings after acquisition.

Pegasystems (Founded in the 1980s, strong in the 2000s)

- Provided workflow automation for CRM, case management, and BPM.
- Focused on rules-driven automation.
- Became a major player in AI-driven process automation later.

OpenText (Metastorm)

- Metastorm BPM was a key workflow automation tool in the early 2000s.
- Provided low-code process automation and enterprise content management.
- Acquired by OpenText in 2011.

BEA WebLogic Integration (Acquired by Oracle in 2008)

- BEA was a leader in Java-based middleware.
- WebLogic Integration was used to automate workflows and integrate apps and B2B transactions.
- Absorbed into Oracle Fusion Middleware.

These companies paved the way for modern workflow automation, robotic process automation (RPA), and AI-driven process management solutions. Many of them were acquired and merged into larger enterprise platforms, while some (like Pega, TIBCO, and K2) continued evolving.

Chapter 7:

Learning from the Past - A Guide for the Future

As demonstrated throughout this book, history doesn't just inform us—it warns us. Every major technological shift—from the rigid structure of the mainframe era to the decentralization of computing via the PC, the rise of the Internet, and the expansive reach of the cloud—has come with remarkable breakthroughs and equally remarkable blind spots. Each phase was heralded as a revolution, and in many ways, it was. But revolutions are often messy, and technology alone has never guaranteed success.

I've witnessed these shifts firsthand. I was there when mainframes ruled the enterprise, and you needed punch cards or terminal sessions to access anything meaningful. When PCs emerged, they didn't just offer power at your fingertips—they disrupted the command structure of IT, introducing decentralization, new risks, and a flood of unsupported applications that IT teams had to chase down. The move to client-server architecture brought scale, but also complexity. The dot-com era introduced agility, but not accountability. And the cloud? It promised freedom from infrastructure, but introduced a whole new dimension of security, governance, and vendor dependency challenges

In every era, the same pattern has repeated itself: we leap toward innovation before we've truly understood what we're building—or how we'll support it. Decisions are rushed, governance is skipped, users are ignored, and leadership often underestimates the long-term needs of the very systems they invest in. We saw it with the uncontrolled sprawl of file shares. We saw it with early intranet portals that nobody used. We saw it with SharePoint, which we'll explore more deeply in a moment. And we're seeing it again today with AI.

If we fail to recognize these patterns, we risk repeating them. The result? Inefficiencies. Wasted investments. Lost data. Burned-out IT professionals. And organizational stagnation at a time when agility is needed more than ever.

Then vs. Now

Mainframe Era → Centralized access (Terminals)

PC Era → Decentralized data chaos

Cloud Era → Centralized again, but with smarter AI tools

AI Era → Requires structured, governed data for automation

And yet, as we move toward this smarter, centralized architecture, shadow IT becomes an even greater threat—not because of malice, but because of impatience. When platforms move too fast, and governance is too rigid, users go rogue. They find the quickest path to productivity: unsanctioned apps, personal cloud accounts, generative AI bots, or free automation tools. The irony? While IT tries to regain control through centralized models, users are scattering again, unlike the PC revolution of the 80s.

The danger here isn't just data leakage, it is fragmentation. Every rogue tool is another island of logic, another place AI can't reach, another blind spot in governance. And the more tools you allow without integration, the harder it becomes to build meaningful automation, security policies, or even accurate analytics. AI doesn't just need access—it needs cohesion.

That is why Tourque was designed as an AI Operating System, not just another app. It assumes this new reality: centralized intelligence, distributed access, and the urgent need to bring structure to chaos. Tourque doesn't fight shadow IT—it integrates around it, absorbs it, and brings visibility to the edges of your enterprise. It treats every process—no matter where it starts—as part of a larger lifecycle. That's what the Process Integration Model (PIM) was built for: to pull rogue tasks and disconnected logic back into a governed, AI-optimized flow without stifling innovation.

We don't need to go backward, we need to evolve forward, learning from the best of the mainframe era but layering it with intelligence, automation, and adaptability. Tourque is that evolution: a secure, structured, and AI-powered platform that doesn't just support the new centralized model—it makes it sustainable.

And here's the most important takeaway: in this new era, the currency isn't computing power or flashy tools, it's structured, trustworthy data. Everything begins with clean metadata, defined governance models, and intentional architecture. Those who embrace this reality will thrive. Those who don't will quickly find themselves locked out of the digital economy's next evolution.

We've been here before. The question is: have we finally learned how to do it right?

Let's ground this in a real-world example that many of us lived through—the promise and pitfalls of SharePoint.

We're currently witnessing a similar wave of excitement around artificial intelligence as we did with SharePoint and other collaborative platforms. New apps and solutions appear daily, many offering impressive features but little in the way of long-term vision or enterprise integration. While some organizations rush to adopt them, many of these tools will not survive. Why? Because the landscape is changing fast. Technology requirements are shifting, hardware is evolving, cloud services are becoming more intelligent and automated, and the demands of AI integration are growing more complex.

This time around, we must be more strategic. Microsoft has long published best practices around metadata, tagging, and content classification—but let's be honest: almost no one followed them. In all the organizations I've worked with, only a handful truly leveraged the SharePoint Term Store or built a sustainable taxonomy. Despite years of documentation, training, and evangelism, most businesses clung to folder structures, ignored metadata, and never implemented the governance needed to support intelligent systems.

But the question we need to ask now is *why*?

Why are we still designing systems around what users want instead of guiding them toward how they *should* be working? Why are we afraid to build structured experiences with enforced best practices that support the long-term health of the organization?

Let's be real: many users didn't *want* to tag documents. They found it inconvenient, unfamiliar, or unnecessary. So, in the name of "ease," organizations removed friction—at the cost of intelligence. But tagging doesn't have to be optional. SharePoint libraries support required metadata fields. Mandatory classification is not a punishment; it's a path to automation. It forces a moment of intentionality. Yet time and again, I've seen these features turned off or ignored, because organizations were afraid of pushbacks.

So again—why?

Is it organizational laziness? A fear of disrupting habits? Or is it something deeper reluctance to admit that we've built digital environments without accountability? That we've prioritized comfort over clarity?

The truth is, most digital strategies stall not because of the technology, but because no one wants to challenge the status quo. Changing management is inconvenient. Governance is hard. Training takes time. So instead of fixing the issues, we kick the can down the road and that road ends right here, in the AI era, where structure is no longer optional.

If we don't correct course now, AI will inherit the same disorganized, half-governed mess we've been avoiding for years. But if we do the work—if we enforce the structures, train the users, and stop treating governance like a dirty word—we'll finally be in a place to build something truly intelligent.

This is the crossroads we've arrived at. For decades, we've allowed habit and hesitation to dictate how we manage content. We avoided enforcing structure, we downplayed governance, and we let user convenience override organizational intelligence. But now, with AI at the center of enterprise strategy, that approach no longer works.

To understand how we got here—and why structure matters more than ever, we need to look at the evolution of technology through a different lens. Not just by devices or software, but by how we've organized information, managed content, and adapted (or failed to adapt) our practices along the way.

The table below breaks down the major technological eras, the prevailing content organization methods, key technological milestones, and the patterns—good and bad—that we must learn from as we enter the AI-driven future:

Years	Technology Era	Content Organization Method	Key Tech Milestones	Notes
1980s–1990s	Mainframes → Early PCs	Folders on local drives	DOS, Windows 3.1, early file systems	User-driven, siloed content
1995–2005	Networked PCs & File Servers	Network drives & folder trees	Windows Server, Novell NetWare,	Shared drives, version chaos

Years	Technology Era	Content Organization Method	Key Tech Milestones	Notes
			Exchange, Outlook	
2001–2010	Intranet Portals & SharePoint	**SharePoint Libraries + Folders**	SharePoint 2003/2007/2010	Basic document libraries with versioning
2010–2015	Cloud & Mobile Era	**Metadata tagging + search**	SharePoint Online, OneDrive for Business	Push toward structured content
2016–2020	Intelligent Cloud & Governance	**Content Types, Policies, Metadata**	Microsoft Graph, Delve, Flow (Power Automate), Purview	Classification begins
2020–Now	AI-First & Zero-Trust Models	**AI-Based Classification**	Microsoft Syntex, Copilot, SharePoint Premium	Content auto-tagging, compliance, security, insights

From Folders to Metadata to AI Classification

Now that AI is here, those decisions—or lack thereof—are catching up with us.

History has already shown us what happens when we rush into technology without planning. From the uncontrolled sprawl of file shares in the 2000s to the disorganized migrations into the cloud, the warning signs were always there. But in no era have the consequences of poor planning been more severe—or more expensive—than in the one we're entering now.

This next phase isn't just another technology cycle, it's a reckoning. AI will not tolerate disorder. It cannot thrive in fragmented, ungoverned, or undocumented environments. It amplifies whatever foundation it's given, meaning chaos will only create more chaos, faster than ever before.

And that's the real threat: not the failure of AI, but the failure to be ready for it. The cost of disorganization is no longer just inefficiency, its lost insights, broken automation, compliance violations, and strategic paralysis. Organizations that ignore the cost of chaos today will pay for it tomorrow in ways they can't yet imagine.

Let's break down exactly what that cost looks like.

The Cost of Chaos

Let me be blunt: the most innovative move your organization can make right now isn't adopting the next flashy AI tool—it's getting your internal house in order.

Because here's what's really happening behind the scenes:

- Shadow IT is running wild.
- Users are driving change, not business.

- They're picking tools based on convenience, not strategy.
- Free apps, automation bots, personal cloud drives, they're everywhere.
- Every unauthorized app is a blind spot.
- Every disconnected platform is a data silo

Every new workaround drags you further from anything resembling a unified AI strategy. Departments are solving the same problems in different ways. Duplicate workflows. Redundant licenses. Tools that do 90% of the same thing, but no one wants to consolidate. Entire budgets are being burned on software that's barely used—or used for one narrow function, another system already does better.

And let's not forget the digital waste.

We're talking about old, outdated documents cluttering every library.
Test flows and half-built apps left abandoned in Power Automate.
Empty Teams sites, broken SharePoint pages, orphaned lists—all from reorgs no one remembers, still indexed, still searchable, and still part of your compliance exposure.

You're paying to store it, to secure it, and to sift through it—every single time someone runs a search, opens Copilot, or tries to automate a process.

This isn't just inefficiency.
This is strategic erosion.

- You can't automate what you don't understand.
- You can't govern what you can't see.

179

- And you can't train AI on a digital junkyard and expect it to build a smart city.

This is the cost of chaos. And it's not invisible—it's sitting right there in your OPEX.
Storage costs. Compliance risks. Wasted licenses.
Missed insights.

That AI project that's running behind schedule? That's not an AI problem—it's a mess problem.

Before you roll out your next chatbot, pilot your next model, or talk about "disruption," start here:

- **Clean your data** – Identify what's valuable, archive what's not.
- **Tag and label content** – Enforce metadata. Use the term store. Don't leave it optional.
- **Audit your tools** – Consolidate duplicate functionality and eliminate shelfware.
- **Archive your trash** – Remove outdated files, test flows, abandoned apps, and digital junk.
- **Document your workflows** – Even if they're messy or manual, they must be understood before automated.

Because AI doesn't fix chaos, it **amplifies** it.

Now let's break down what that chaos is really costing you.

You can't fix what you're not willing to face. So, let's put some numbers to the mess—and show exactly what it's costing you.

The Cost of Chaos: A Real-World Breakdown

This table highlights the hidden (and often ignored) financial and operational tolls of fragmented systems, outdated content, and poor governance.

Category	Hidden Cost	Real-World Impact
Shadow IT	$50k–$250k/year	Licensing and security exposure from unsanctioned apps
Duplicate Tools/Platforms	$100k+/year	Paying for redundant software with overlapping functions
Unused Licenses	$20–$50/user/month	Assigned tools not being used by staff
Outdated Content & Junk Data	Up to $190k/month in storage	Cloud storage overages from old/test/irrelevant files
Disconnected Workflows	Lost productivity (hours/week)	Teams redoing work due to lack of centralized processes
Compliance Risks	$100k+ fines per incident	Sensitive files unlabeled or exposed due to poor governance
AI Rework Costs	$$$ in consulting and delays	AI models failing due to

Category	Hidden Cost	Real-World Impact
		fragmented or unstructured data

Shooting in the Dark: The Real Cost of Rushing into AI

The AI hype cycle is in full swing and with it comes a dangerous wave of premature enterprise adoption. Companies are eager to be seen as "innovative," but few are truly prepared. They are signing high-cost contracts, deploying fragmented pilots, and spinning up isolated agents—often without fully understanding the burden they're taking on.

Here is the reality:

- AI pricing models today—especially for conversational tools like Copilot—are steep. Per-user or per-agent pricing adds up fast, especially when applied to project work, not just customer service.
- Early adopters are locking into systems that may not scale, or flex as better tools emerge. And once your processes are built around one tool, migrating later is both risky and expensive.
- Most organizations still have broken, siloed processes underneath. AI will not solve that. In fact, it makes disjointed workflows even more apparent—and more frustrating.
- The promised ROI rarely materializes because governance, structure, and true operational integration were never addressed.

In short: companies are shooting in the dark, desperate to move forward, but there is no clear path ahead—just a lot of expensive noise.

That is exactly why I built Tourque.

Tourque is not just another AI add-on. It is an AI Operating System—designed to bring structure, scalability, and sanity to enterprise AI. With Tourque, you do not just automate tasks, you evolve how your organization *thinks*, *learns*, and *operates*.

The real AI transformation is not about getting ahead of the curve—it's about building a curve that leads somewhere.

Chaos Symptoms vs. Strategic Fixes

The symptoms might look like day-to-day "normal" work—but they are hiding serious inefficiencies. This table pairs common organizational behaviors with their hidden costs and practical solutions.

Symptom of Chaos	What It is Costing You	Fix
Users hoarding files "just in case"	Ballooning storage and slower search performance	Implement retention policies
Abandoned Power Automate flows/apps	Confusion, security risk, broken processes	Audit and clean up test environments
Rogue Teams using 3rd-party AI tools	Data leakage, misalignment, compliance gaps	Centralize AI strategy & tool selection

Symptom of Chaos	What It is Costing You	Fix
Same process built 4 separate ways	Wasted time, rework, inconsistent output	Adopt a unified workflow architecture
Files with no tags or versioning	AI failure, poor discovery, compliance risk	Enforce metadata and version control

Quick Audit: Are You Paying for This?

Here is a rapid-fire checklist to help you spot chaos hiding in plain sight. If you are answering "yes" to more than a few of these, you're bleeding budget and burning time.

Question	If "Yes," You're Paying For...
Are users using apps that IT does not support or monitor?	Shadow IT, security gaps
Do multiple departments use different tools for the same task?	Redundant licensing, inconsistent workflows
Are you storing files that have not been touched in 5+ years?	Wasted storage, indexing lag, AI drag
Do you have test workflows and apps still deployed in production?	Governance risk, confusion, clutter
Are your AI projects running slower than expected?	Poor data quality, fragmented inputs

Data is one thing but lived experience tells the story even better. I have seen firsthand how costly digital disorder can be—not just in dollars, but in time, morale, and missed opportunities.

I recently worked with a company so far behind on content governance that it would take them an estimated five years just to clean up their SharePoint environment before they could even begin exploring AI. And to be clear—it wasn't just about them dragging their feet. The tools themselves were bottlenecks. Running content discovery scripts in Microsoft 365 takes time and compute. I once ran a PowerShell script to pull metadata from a single (small in their environment) site collection, it took three days to complete.

Even basic tagging operations can be painfully slow. Tagging a document using Microsoft Purview or other cloud classification tools can take up to 48 hours to process across a SharePoint environment. That's not just inconvenient, it's a major roadblock when you're trying to secure sensitive files, apply governance policies, or prepare your data for AI-driven automation.

Combine that with the time it takes to configure compliance centers, assign roles, train users, and coordinate across departments, and suddenly, your AI roadmap isn't delayed, it's stalled.

If your content is a mess, your AI will be too. But if your foundation is strong, you're not just prepared for the AI era, you're ready to lead it.

Speaking of leading it, one of the most dangerous contributors to digital chaos isn't bad technology or lazy users, it's leadership pressure without planning.

In many organizations, the C-suite is demanding progress—fast. They want AI now. They want automation by next quarter. They want to be able to tell shareholders, boards, and partners that their company is "AI-powered," "cloud-first," or "digitally transformed."

But here's the truth: urgency doesn't equal readiness. And shouting "go" doesn't build a roadmap.

I've worked with countless organizations where managers and technical leaders were given a mandate with zero strategy behind it. "Get this AI chatbot online," or "Migrate everything to SharePoint this quarter," or "Set up Power Automate flows to speed up approvals." These aren't strategies, their orders without context. And they almost always lead to shortcuts, patch jobs, and surface-level results.

It's not that the C-suite is wrong to push for progress. But progress without planning is just pressure, and pressure without support leads to brittle, rushed solutions that look good on a slide deck but fall apart in practice.

Here's what gets missed:

- The **technical debt** of moving too fast
- The **operational chaos** from skipping discovery
- The **burnout** of IT teams working under unrealistic timelines
- The **fragility** of systems cobbled together to meet an arbitrary launch date

Too often, success is declared just because something is "working." But just because a system functions on the surface doesn't mean it's scalable, secure, or sustainable.

The organizations that thrived in the AI era will be the ones that slow down just enough to do it right:

- Who gives their technical teams time to plan, structure, and test.
- Who invests in proper documentation and content cleanup.
- Who trust their experts and prioritize long-term stability over short-term optics.

Executives need to understand: AI isn't a feature—it's a framework. And frameworks require architecture. If you don't lay a strong foundation, you're not building the future—you're stacking cards in a windstorm.

C-Suite Manifesto: What Leaders Need to Hear About AI Readiness

Patience leads to performance
Give your teams room to build it right the first time.

Governance is not optional
No amount of innovation justifies disorganized or insecure systems.

You can't fix what you rush past
Slow down now—or clean up later at 10x the cost.

Your tech team needs more than orders
They need strategy, clarity, and cover from the top.

Without that foundation, you risk repeating the mistakes of the past. With it, you can genuinely position your organization to harness the power of AI—not just today but in the long run. The hype cycle is real, but so are the opportunities. Don't fall for the flash. Focus on the fundamentals.

The AI Gold Rush

If the cost of chaos wasn't enough, we now have a new layer of risk: **the AI gold rush.**

Much like the dot-com bubble or the SharePoint boom, organizations are once again racing to adopt the latest this time, it's AI. And just like those eras, the excitement is drowning out the strategy.

We've seen this story play out before, not just with AI, but with Agile, ITIL, and every framework that promised order but failed to deliver when execution fell apart. Let me share one such story from the field.

Boards want AI dashboards. Executives want AI pilots. Teams are hiring AI consultants with slick slides but no hands-on experience. It's not transformation—it's theater.

I've seen companies jump headfirst into AI tools without:

- A clear use case
- Clean data
- A content strategy
- Or even the right personnel to support it

It's déjà vu. We're watching the same mistakes unfold—just wrapped in a shinier package.

So, let's break this down. There are three critical lessons we *should* have learned by now:

1. **AI is a tool, not a solution.**
   It won't fix broken processes; it only automates what's already there. If your workflows are chaotic, AI will make them faster... and worse.
2. **Structured data is non-negotiable.**
   If your content is untagged, fragmented, or outdated, AI can't find patterns. It can't classify, retrieve, or generate accurate results.
3. **People still matter.**
   AI doesn't run itself. Without proper training and skilled talent, your AI investment will stall, misfire, or backfire.

This isn't about being first, it's about being *ready*. In the gold rush, not everyone finds gold. Most just dig holes.

The problem isn't that organizations are adopting AI—it's that they're doing it without fixing what's broken underneath. What looks like progress on the surface is often hiding a tangled web of outdated systems, redundant tools, unstructured content, and inconsistent processes.

Before you go any further, look at what's really going on behind the scenes. This is the cost of chaos visualized.

The Cost of Chaos

Visible and Hidden Problems

The Cost of Chaos: When Structure Is Ignored, Efficiency Pays the Price

Disconnected systems, duplicated data, shadow IT, and ungoverned tools don't just slow you down—they compound cost, risk, and missed opportunity. The real price of chaos isn't technical, it's strategic. Organizations can no longer afford to repeat the mistakes of past technology shifts. Top tech companies have been laying groundwork for years, introducing best practices, promoting digital maturity, and urging enterprises to prepare for what's next. But after working with over 130 companies throughout my career, I can tell you this: very few listened.

The result?
Missed opportunities.
Mounting technical debt.
And now totally unpreparedness for AI.

If we had followed even half the guidance from a decade ago—on metadata, governance, taxonomy, lifecycle management—we'd be stronger today. But we didn't. And now we're entering an AI-driven world, still dragging around bad habits and bloated systems.

AI isn't just about excitement or experimentation.
It's not a trend.
It's a fundamental shift in how work gets done—and it demands strategic foresight, clean data, consistent governance, and real integration.

The companies that understand this—and act—will lead the future.
The rest?
They'll spend the next five years cleaning up yesterday's mess, trying to catch up—if they even can.

We've seen this play out before. During the SharePoint wave. During cloud migrations. The warning signs were there. But many failed to plan, and what followed was predictable: disillusioned users, fragmented platforms, shadow IT, and failed rollouts. Now, with AI, those same dangers are not only back— they're bigger.

Because this time, the tech is moving faster than the organizations trying to adopt it.

This is the moment.
Either we fix the foundation—or we watch the next revolution pass us by.

These aren't new lessons. We've been here before—with mainframes, PCs, the web, SharePoint, and cloud. AI is just the next wave—but the fundamentals haven't changed.

- **Governance isn't optional.**
  Unstructured, unmanaged content will always brake systems—AI just makes it more obvious.
- **Metadata is power.**
  Search, security, and automation all fail without classification and tagging.
- **You can't shortcut strategy.**
  Skipping discovery, planning, and documentation guarantees future failure.
- **Training is everything.**
  Technology adoption dies when users are confused, unsupported, or left behind.
- **Shadow IT is a symptom, not a cause.**
  When platforms don't serve the users, users find their own tools—and chaos follows.
- **Tools don't solve problems, processes do.**
  The best tech in the world can't fix a broken workflow or a siloed culture.

I remember working with one organization where we wanted to stress test a SharePoint environment by loading it with real-world content. To simulate a live, heavy-use site, I built a PowerShell script that automatically populated document libraries and document sets across the site. But we didn't just dump files, we did it right.

We used managed metadata to tag every document on the site. Why? Because we weren't just testing performance—we were testing governance. We wanted to make sure that when it was time to reuse the site, we could reset everything quickly and cleanly, without any manual cleanup.

And that's exactly what we did.

With a single PowerShell command filtered by the metadata tag I had assigned, I wiped the site clean in minutes—no broken pages, no ghost files, no hunting through folders. That's the power of structured content. That's the payoff of doing the groundwork. Imagine the reporting you could create at your fingertips if everything was labeled and structured.

This kind of planning isn't just about good hygiene— it's about flexibility, speed, and control.
And unfortunately, it's the kind of discipline most organizations skip—only to realize later how badly they needed it.

There are other areas of an organization's foundation that need to be grounded to see efficiency effectively.

Structured Testing with Metadata: A Real-World Example

> **6 Reuse the Site**
> Start fresh without manual cleanup

> **5 Clean with Precision**
> Use PowerShell + tags to wipe site

> **4 Simulate Usage**
> Test performance and usability at scale

> **3 Tag with Metadata**
> Apply managed metadata to eachument

> **2 Automate Upload**
> Use PowerShell to populate docunds

> **1 Plan the Test**
> Define objectives for load and cleanup

Structured Testing with Metadata: A Repeatable Testing Framework

This visual outlines a six-step process for testing document environments using metadata and automation. By leveraging PowerShell and managed tagging, teams can simulate usage, validate performance, and reuse environments efficiently without manual cleanup.

Case Study: When Agile Collapsed and SharePoint Suffered

Project Context:

I worked at a large enterprise that began preparing for a SharePoint Online migration nearly two years before the final cutover deadline. The early momentum was promising user stories were being documented, managed metadata was being planned, and the goal was to clear use SharePoint properly to bring order and structure to multiple teams.

The Breakdown:

But within months, everything stalled. I was laid off during the planning phase, and without me, progress froze. My only teammate who held key knowledge of the legacy SharePoint system was also gone. Team leaders weren't engaged. Most were distracted by other work, including the development of custom-coded SharePoint applications, despite the new environment explicitly forbidding custom code. I raised this issue before I left, but the warnings went ignored. Development on a massive app—dozens of screens and countless lists—continued anyway.

The Agile Illusion:

Agile, in theory, should've helped. But in practice, it caused more friction. Every sprint requires a full week just for planning. As a developer working on complex solutions, losing four days of focus on planning discussions didn't help me move anything forward. The rhythm of Agile became a distraction. We weren't building, we were talking. It was a distraction.

The Comeback (and the Chaos):

I was brought back six months later as a contractor. But by the time we got aligned, we were down to only a few months before the mandated migration date. We asked for more time and were granted an additional two months. Still, getting everyone aligned internally took nearly three more months. To help, internal resources were brought in but they had zero context. Onboarding and approvals chewed up more weeks. By the time we were truly ready to start, only two months remained.

What Did We Do?

We dropped Agile completely. There wasn't time to iterate or plan—we had to deliver. I built and deployed five Power Apps in record time. But even then, I still never received the access I needed to Power Automate flows despite waiting over seven months for clarification. The service catalog process was broken from the start.

Lessons Learned:

- **Agile is not guaranteed,** especially when leadership isn't aligned, and focus is constantly broken.
- **Planning means nothing without execution.** The best sprint board won't matter if no one's doing the work.
- **Custom code without governance is a ticking time bomb.** Building for now while ignoring what's next creates migration blockers and technical debt.
- **Frameworks like Agile or ITIL can't fix cultural dysfunction.** Execution, ownership, and unified purpose are what move teams forward.
- **Enterprise work requires flexibility—not just process.** What saved this migration wasn't Agile, it was adaptability, direct ownership, and a willingness to shift when time demanded it.

So where do we go from here?

If your foundation is shaky, AI won't help you, it'll expose the cracks.

Let's talk about what real AI readiness looks like.

AI Readiness: You Can't Fake the Foundation
Microsoft has been promoting structured content for over a decade—metadata, tagging, classification, and governance have been well-documented. But let's be honest: most organizations ignored it.

Now, with AI technologies like Microsoft Syntex and Copilot gaining traction, that oversight is catching up with them. The companies that treated metadata as optional are now discovering that unstructured content is a wall—not a runway.

AI is only as good as the data it touches. If your environment is filled with outdated, inconsistent, or poorly organized content, your AI adoption will be slow, frustrating, and expensive. You won't get meaningful results, you'll get noise.

To get ahead, organizations need to take a proactive approach:
- **Conduct a full content audit** – Know what you have, and more importantly, what you *don't* need.
- **Clean up ROT (redundant, obsolete, trivial) content** – Archive or delete what doesn't serve a purpose.
- **Structure your data** – Use metadata, taxonomy, and tagging to make content discoverable and trainable.

- **Work with legal and compliance teams** – Make sure your AI strategy aligns with governance and retention policies.
- **Define clear use cases** – Don't chase shiny tools—build intentional, outcome-driven AI solutions.

This isn't just about optimizing content, it's about preparing your digital ecosystem for automation, insight, and scale. AI is not forgiving. It won't sort through your mess for you. But if you give it structure, it will do heavy lifting.

Top 5 Excuses I've Heard from Companies Not Ready for AI:

1. **"We'll just clean it up later."**
   → You won't. And by then, your AI project will already be behind.
2. **"Users don't want to tag documents."**
   → It's not about what users want, it's about what the business needs.
3. **"We don't have time to audit our content."**
   → Then you don't have time for AI either.
4. **"We already migrated to the clouds we're good."**
   → Moving junk from one place to another is not transformation.
5. **"We'll hire an AI vendor to figure it out."**
   → They can't fix your foundation. They'll either overcharge or underdeliver.
6.

Comparative Frameworks

We've talked about AI readiness at the organizational level—about cleaning up your data, enforcing governance, and building a strong foundation. But this isn't just about internal operations. It's about positioning your entire company to survive the next technological shift.

History has already shown us what happens to organizations that resist change or delay transformation. The rise and fall of industry giants follow a predictable pattern—and it's playing out again in the AI era.

AI isn't just a new capability—it's a new competitive filter. The companies that move first *and* execute well will dominate. Those who rush blindly—or wait too long—will fall behind.

We've seen it before:

- **Microsoft** adapted through every major wave— from desktop to enterprise to cloud—and continues to thrive.
- **IBM**, once dominant, failed to pivot fast enough in the personal computing era and slowly lost relevance in the consumer space.
- **Apple** didn't just react to mobile—it reshaped the world around it.
- **BlackBerry** and **Sears** stuck to what used to work—and disappeared from the front lines of innovation.

Tech Titans That Lost Their Edge: Lessons in Adaptability

Sony PSP: A Missed Opportunity in Mobile Gaming

The PlayStation Portable (PSP) was a technological marvel at its launch, offering console-quality graphics in a handheld device. However, several factors led to its decline:

199

- **Proprietary UMD Format**: The reliance on Universal Media Discs (UMDs) limited game accessibility and was less convenient compared to cartridges or digital downloads.
- **High Piracy Rates**: The PSP was easily hackable, leading to widespread piracy that deterred developers from investing in the platform.
- **Competition from Nintendo DS**: The DS offered a more extensive game library and innovative features like dual screens and touch input.
- **Rise of Smartphones**: The emergence of smartphones provided consumers with multifunctional devices, reducing the appeal of dedicated gaming handhelds.

Yahoo: A Case of Missed Opportunities

Once a dominant force in internet services, Yahoo's decline was marked by several strategic missteps:

- **Failed Acquisitions**: Yahoo missed the chance to acquire Google in its early days and later failed to purchase Facebook.
- **Lack of Innovation**: The company struggled to innovate and adapt to the rapidly changing digital landscape.
- **Security Breaches**: Massive data breaches eroded user trust and tarnished the brand's reputation.
- **Leadership Challenges**: Frequent changes in leadership led to inconsistent strategies and a lack of clear direction.

Nextel: Innovation Without Adaptation

Nextel's push-to-talk feature was revolutionary, catering to industries like construction and logistics. However, the company's downfall was precipitated by:

- **Merger Missteps**: The merger with Sprint led to cultural clashes and integration issues, diluting Nextel's unique value proposition.
- **Technological Lag**: Nextel failed to transition effectively to newer technologies like 3G and 4G, falling behind competitors.
- **Customer Service Issues**: Post-merger, customer satisfaction declined due to service disruptions and poor support.

MySpace: The Social Media Pioneer That Couldn't Keep Up

MySpace was a trailblazer in social networking but eventually lost ground due to:

- **User Experience Challenges**: Cluttered interfaces and slow load times frustrated users.
- **Over-Monetization**: An aggressive focus on advertising compromised the user experience.
- **Security Concerns**: Issues like cyberbullying and data breaches raised safety concerns.
- **Failure to Innovate**: MySpace couldn't match the streamlined experience and continuous innovation offered by emerging platforms like Facebook.

Lessons Earned: The Imperative of Vigilance and Agility

These narratives underscore a critical truth: lessons learned are lessons earned. The downfall of these tech giants wasn't due to a lack of innovation at inception but rather a failure to adapt and evolve. In today's AI-driven economy, the competitive edge lies not just in groundbreaking technology but in operational efficiency, customer-centricity, and data utilization.

- **Efficiency Over Size**: Streamlined operations can outperform larger, less agile competitors.
- **Customer Experience as a Differentiator**: Rapid, personalized service fosters loyalty and sets companies apart.
- **Data as a Strategic Asset**: Clean, actionable data is the foundation for effective AI implementation and informed decision-making.

In this landscape, complacency is a liability. Companies must remain vigilant, continuously assess the competitive environment, and be prepared to pivot strategies in response to emerging trends and technologies.

Blackberry: Brilliant Infrastructure, Disrupted by the Internet Boom

BlackBerry once led the mobile communication market with innovative devices and infrastructure. They persuaded enterprises to install BlackBerry Enterprise Servers (BES) for secure email and messaging. This offered unmatched mobile security, integration with Microsoft Exchange, and device management.

But innovation doesn't happen in a vacuum. While BlackBerry was doubling down on their tightly controlled, server-driven ecosystem, the public internet was rapidly expanding.

- **High-speed mobile internet (3G, then 4G)** opened the mobile web to everyone.
- **Apple's iPhone (2007)** and the **rise of Android devices** removed the need for company-specific servers and infrastructure.
- Suddenly, users could access email, apps, and web services directly, without needing a BlackBerry-specific ecosystem.

BlackBerry had solved the right problem—too late for the new world that was emerging.

The infrastructure they had so carefully deployed—servers in almost every major corporation—became a liability instead of an advantage. Companies didn't want to manage on-premises servers anymore; they wanted cloud services, simplicity, and devices that could integrate natively with everything.

Key Lessons from the BlackBerry Story

- **Infrastructure Lock-In is Risky:** Building your value around proprietary infrastructure can backfire when open standards or cloud solutions emerge.
- **Market Timing Matters:** Innovating early is important—but staying flexible as the environment changes is even more critical.
- **User Experience Wins:** Convenience often beats security and control. When smartphones made email and browsing easy without extra servers, the market shifted overnight.

The same story is unfolding with AI. The winners of this era will not be those who *experiment the fastest—* but those who *prepare the smartest.* That means:

- Investing in the right people—AI leadership, not just headcount
- Understanding the real risk bias, hallucination, regulatory gaps
- Embracing governance and integration, not chasing quick wins

BlackBerry — A Hidden Lesson on Infrastructure Risk

Visionary Move:

BlackBerry placed secure servers (BES) inside corporate data centers worldwide, revolutionizing executive mobile communication.

The Missed Shift:

With high-speed mobile internet and smartphones eliminating the need for private infrastructure, BlackBerry's advantage turned into a disadvantage.

Companies moved toward the cloud. Flexibility, not control, won.

The Hard Lesson:

- Solving yesterday's problems can leave you vulnerable tomorrow.
- Rigid infrastructure strategies don't survive market shifts.
- Convenience and adaptability often outweigh technical superiority.

What It Means for the Future

In the AI era, the same warning applies:

Don't build systems that trap you. Build systems that move with you.

Infrastructure Risk in the Age of AI: A Critical Wake-Up Call

As we step deeper into the AI era, one truth becomes unavoidable:

- Your infrastructure choices today will either empower you—or expose you—tomorrow.
- The cloud itself is evolving.
- New architecture, new security challenges, new processing models.
- Choosing the wrong partners, building in rigid environments, or locking into yesterday's technologies could cripple organizations faster than any past tech wave.

That's why alignment matters more than ever.

You don't just need cloud partners; you need vision partners.

Partners who are evolving toward automation, AI-native architectures, hybrid cloud flexibility, decentralized models like blockchain, and real-time data governance.

Tourque is part of that disruption. But it won't stop with Tourque. There will be other new platforms, new models of collaboration, new ways of securing and governing information.

You want to ride *that* wave of innovation—not be dragged by it.

Walk aimlessly through changing times or chart your course with AI-driven insights.

One critical advantage we now have:

AI itself can guide us.

Today's AI tools can analyze market trends, flag emerging technologies, detect shifts in customer behavior, and even predict where disruption is likely to occur.

If you're willing to listen, AI can show you where the world is heading.

Key Takeaways:

- Infrastructure agility is survival.
- Alignment with future-focused platforms is essential.
- AI isn't just a tool—it's a compass for business strategy.
- Stay flexible, stay aware, and be willing to pivot when the signs appear.

Mindfulness, intentionality, and humility will define the winners of this next era.

The companies that fail to prepare for AI aren't just risking a bad rollout; they risk risking irrelevance.

The Cost of Ignoring the Past

The technological revolution is underway. AI, automation, and machine learning will reshape industries, redefine jobs, and create new economic models.

History has shown what happens when companies fail to prepare. The dot-com bubble revealed the dangers of overhyped investment in unproven technology. The early internet boom demonstrated that speed of adoption matters, but execution matters more. The SharePoint era showed that adoption without governance leads to failure.

If AI is not strategically implemented, history will repeat itself once again.

Those who prepare will thrive. Those who rush in without strategy will fail. Those who ignore AI entirely will be left behind.

The Price of Neglecting Governance

I've seen a handful of truly mature environments in my career—where the term store was properly organized, Power Automate flows enforced retention policies, and SharePoint wasn't just a dumping ground but a *governed platform with purpose*. But here's the truth: those environments are rare.

Governance and structure have been Microsoft's message since SharePoint 2007, but too few have truly listened. Most organizations are sprinting just to keep fixing fires, launching new tools, and chasing productivity. But in doing so, they skip the very foundation that enables long-term success: clean data, thoughtful governance, and reusable architecture.

Instead of asking, "How can we improve our SharePoint or document environment?" the focus becomes, "What's the next solution we can spin up?" That mindset creates silos, duplicate systems, and fragmented data.

You end up with:

- Multiple databases storing the same information.
- Departments are running their own disconnected reports.
- Redundant apps doing the same thing with different names.
- And licensing waste from duplicated, unsanctioned platforms

It does not have to be this way.

When does it stop? It stops when leadership shifts its mindset—when we start prioritizing structure over speed, clarity over convenience, and governance over gimmicks.

AI won't fix chaos.
AI will amplify whatever structure—or disorder—you already have.

If your environment isn't built like a well-oiled machine, it will act like the Wild-Wild West.

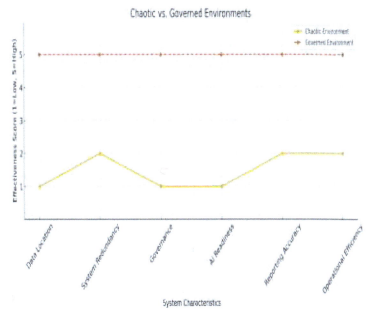

Chaotic vs. Governed Environments

Preparing for the Future: The Call to Action

Organizations must audit their technology environments to ensure AI success to determine if their data is AI-ready. Investing in AI education and hiring experts with real-world experience is crucial. AI governance models must be built to define how AI will enhance rather than disrupt business operations. Companies should adopt AI for the right reasons rather than simply following competitors. Above all, organizations must commit to continuous learning and adaptation.

What should companies be doing right now?

- **Audit your technology stack.**
  Understand where your data lives, how it's structured, who owns it, and whether it's usable by AI systems.
- **Invest in AI education.**
  Not just for IT leaders—but for legal, compliance, HR, operations, and business units. AI is a cross-functional transformation.

- **Hire people who've done the work.**
  Not just consultants with theory—but people
  who've built real-world automation and
  governance systems.
- **Build an AI governance model.**
  Define how AI will be used, how data will be
  secured, how decisions will be audited, and
  who's accountable for outcomes.
- **Move with purpose, do not panic.**
  Don't adopt AI just because your competitors
  are. Define clear, measurable outcomes.

Above all, commit to continuous learning and
iteration. AI will change everything—but only for those
who are ready to evolve alongside it.

AI will change everything. The question is, who will be
ready?

Why Cloud Migrations Were Critical for AI Transformation

In the rush toward digital transformation, many
organizations saw Microsoft 365 as the obvious next
step. Cloud-based platforms promised scale,
automation, and flexibility—key ingredients for an AI-
powered future.

But there was a problem: most migrations were
tactical, not strategic.
They moved data but didn't clean it.
They shifted storage but didn't rethink structure.
They changed location but didn't transform behavior.

For many, these migrations created more problems
than they solved:

- Legacy content was dumped into modern
  platforms without tags or retention.

- Permissions and access policies were inherited—unreviewed and unoptimized.
- Cloud costs ballooned because no one filtered out junk before the move.

Suddenly, the cloud wasn't just a new home, it was a new mess.

For many, migrations were seen as a quick way to move away from on-premises systems, but without careful planning, they created storage chaos, compliance risks, and performance bottlenecks. The slow processing speed of Azure Information Protection (AIP) or Microsoft Purview labeling is a clear example of the consequences of rushed migrations. With a 48-hour delay in processing labels, organizations were exposed to security risks and compliance gaps.

Why This Delay Is a Problem

1. Legal Holds & Sensitivity Labels: AI-driven automation relies on accurate and immediate document classification. Organizations face potential compliance violations and data breaches if sensitive files remain unprotected due to labeling delays.
2. Storage Management Issues: Rushed migrations resulted in millions of unnecessary files being moved—old documents, outdated versions, personal media, and non-essential data. The consequence? Massive storage bloat made AI-driven processing inefficient.
3. Data Leakage & Unauthorized Access: When labeling speeds lag, documents remain vulnerable. This means files can be moved, altered, or deleted before properly securing an open invitation for security risks.

Are We Sacrificing Agility for Cloud Adoption?

Organizations moved to the cloud for scalability and flexibility, yet the processing delays, lack of governance, and mismanaged storage are slowing businesses down rather than making them more efficient. The key takeaway? If your data is poorly structured, your AI roadmap is already off-course.

We moved to the cloud for speed, scalability, and innovation. But what many companies got was lag, confusion, and bloated environments.

The irony? In trying to become more agile, we lost clarity.

We traded physical servers for virtual sprawl. We gained scalability but lost structure.

Processing delays, misconfigured governance centers, and endless storage silos now bog down AI initiatives that should be running at speed. Instead of a streamlined platform, we're managing layers of complexity that were never cleaned up post-migration.

The question is no longer "Should we move to the cloud?"

It's "Did we move the right way—and did we clean up after ourselves?"

Because if your data is disorganized, your AI roadmap isn't just off-track—it's in the wrong direction entirely.

Post Migration Reality Check

Throughout my career, I've seen one consistent failure across organizations of all sizes: they never truly embraced metadata. Even when SharePoint first emerged as a structured content platform, most site administrators defaulted on folders instead of leveraging the Term Store and content types. That decision had consequences.

Metadata wasn't just a nice-to-have. It was always essential to intelligent systems—and now with tools like Microsoft Syntex, it's become a requirement. Without structured metadata, AI lacks context. It can't categorize content, process automation, or deliver meaningful insights. The companies that assumed "we'll fix it later" are now finding themselves blocked from fully adopting AI at all.

I worked with one of the largest SharePoint environments I've ever seen, an unstructured migration dumped everything into SharePoint Online without curation. File shares, outdated content, non-collaborative documents, even executable files—were blindly moved to the cloud.

The result?

- **Storage bloat**: 100's of Site collections ballooned to 25TB, maxing out limits.
- **Content chaos**: 40% or more of the data had no business value—MP3s, ZIPs, old backups.
- **Governance bottlenecks**: Cleanup scripts took days. Tagging operations took 48 hours just to label content.

This wasn't modernization. It was digital hoarding.

The Illusion of Collaboration

Microsoft 365 introduced powerful collaboration tools—co-authoring, synchronizing, version control but too many organizations rolled them out without governance. They assumed collaboration would "just happen." Instead, it led to:

- Disorganized document libraries
- Confusion from multiple versions of the same files
- No permissions strategy, no retention plan, and no training

AI can't work in this kind of environment. Without structure, automation becomes error prone. Without clean data, AI insights are inaccurate. And without governance, collaboration turns into chaos.

Quick wins don't build long-term value. Many organizations believed moving to the cloud or adding AI tools would bring instant results. But without clean data, structured content, and proper oversight, they've only accelerated their own confusion.

So, what should companies be doing now?

What You Need to Do Now

1. **Audit Your Data**: Remove outdated, irrelevant content.
2. **Prioritize Metadata**: Tag content before AI gets involved.
3. **Rethink Governance**: Align automation tools with real compliance strategies.
4. **Train Your Teams**: If users don't understand metadata or workflows, AI won't help.

Because the truth is: AI doesn't care about your data, but your company should.

Post-Migration: Did You Achieve What You Promised?

Many companies moved to Microsoft 365 with goals like cost savings, scalability, collaboration, and modernized systems. But now they're asking, *"did we get there?*

Post-Migration Reality Check: Are You Really AI-Ready?

Expectation	Reality	AI Impact
We moved to the cloud to modernize	Legacy folder structures and junk data came with it	AI can't categorize or automate effectively
We thought we'd reduce costs	Unfiltered data is driving massive storage and licensing fees	Budget is bloated, not optimized
Collaboration would improve	Users abandoned tools or created chaos without training	Disconnected workflows and shadow IT
We planned to scale easily	Poor architecture created performance issues and silos	AI performance is slow or unreliable
AI will unlock value automatically	Content is untagged,	AI results are inaccurate or unusable

Expectation	Reality	AI Impact
	unmanaged, and hard to interpret	

If this sounds familiar, your AI tools won't deliver until your foundation is fixed.

Cloud Migration Promises vs. Reality

Cost Savings
The cloud was supposed to reduce infrastructure costs. But in practice:

- Bloated storage costs from unfiltered migrations
- Hidden licensing fees, VDI complexity, and new operational overhead
- One company I worked with spent over 100K/month on unnecessary cloud storage

Scalability
Cloud platforms offer scale—but poor architecture and vendor lock-in now limit flexibility.

Collaboration
Despite access to cutting-edge tools, most employees still feel disconnected. Why?

- Too many overlapping platforms
- No unified strategy or training
- Shadow IT fills the gaps, introducing risk

Fragmentation & Shadow IT
The promise of collaboration tools was clear, but the reality is far more fragmented. Collaboration was expected to improve teamwork, yet studies show that 70% of employees still feel left out of decision-making processes (Gartner). Why?

- Too many disconnected tools create confusion rather than efficiency.
- Lack of proper training results in underutilized platforms.

And nowhere is that fragmentation more dangerous than in the age of AI—where even small gaps in alignment can derail enterprise-wide automation.

At one company I worked with, I was hired to conduct interviews with several teams that were supposed to use SharePoint as their primary collaboration platform. The goal was to assess adoption and identify any roadblocks.

What I discovered was surprising: nearly 90% of users had completely abandoned SharePoint. When I asked why, the responses were consistent:

- "I don't know how to use it"
- "I've only heard bad things about it"
- "It's too complicated to figure out"
- "We never received any training"

Digging deeper, I found the issue was more widespread than expected. It wasn't just one or two departments—close to 90% of the company's teams had quietly turned to their own preferred tools to get work done. These included open-source platforms, free apps, and cloud-based services that weren't sanctioned or visible to the organization's IT department.

The reason? Familiarity. People chose tools they already understood, even if they weren't secure, supported, or integrated into the company's ecosystem.

The result was a fragmented environment filled with shadow IT, unsecured data, and a complete breakdown in visibility and governance.

We cannot afford to repeat this mistake with AI

If we let people adopt AI tools in isolation—without governance, integration, or purpose—we risk creating an even greater mess than before. Unlike past technologies, AI feeds on data, and if that data lives in disconnected, ungoverned apps, the insights we expect from AI will never come. Worse, we'll create compliance risks and lose control of valuable business knowledge.

I was at a meeting where an organization announced that its enterprise AI offering was nearing release. They were still finalizing pricing and rollout schedules. Yet, the next day, someone posted in the company chat that they were already aware of more than a hand full of other AI implementations and wanted to explore integrating them into the new, still-unreleased enterprise AI solution.

This is precisely what I mean when I say governance is key. If you don't get a handle on your users and developers early, you'll face challenges around adoption, fragmented data, and alignment. When rogue teams move ahead without coordination no matter how well-intentioned they create a fractured ecosystem that undermines enterprise unity. What starts as innovation can quickly turn into redundancy, wasted hours, and technical debt.

Unchecked Shadow IT = Fragmented AI

When every team builds their own AI stack, you're not enabling innovation, you're breaking the foundation before it's poured.

Now, a subset of users is invested in tools that may never align with the approved solution. Where does that leave the rest of the organization? Behind, misaligned, and spending more time cleaning up than building forward. This is not just a detour, it's a derailment. And that's why strong governance must come before the rollout, not after.

Before & After: AI Rollout Without vs. With Governance

Category	Before (Shadow IT Chaos)	After (Governed AI Readiness)
Tool Selection	Teams pick their own tools based on convenience	Tools chosen based on enterprise strategy
User Behavior	Adoption driven by familiarity, not alignment	Guided adoption with training and support
Data Structure	Scattered, inconsistent, and untagged content	Structured, unified, and metadata-rich
Governance	No central oversight or process control	Governance frameworks established and enforced
AI Integration	Isolated pilots and	Cohesive enterprise AI

Category	Before (Shadow IT Chaos)	After (Governed AI Readiness)
	fragmented models	roadmap with phased implementation
Security & Compliance	Risks untracked, data lives in unsecured systems	Security controls and compliance built into every layer

Security, Simplicity, Governance

The convenience of cloud collaboration has come with a cost: increased security vulnerabilities. While platforms today make it easier than ever to share files and collaborate across teams, that same ease has introduced risks like intellectual property leaks, misconfigured permissions, and uncontrolled data access.

According to a study by the Ponemon Institute, human error is responsible for 23% of all data breaches, a sobering reminder that even the most advanced systems can be brought down by something as simple as a mis click. This reinforces a critical point: Security must be designed for the user, not against them.

Too often, organizations rely on exhaustive—and frequently confusing—security training programs that overwhelm employees with jargon, technical details, and complex workflows. While intentions may be good, this approach doesn't always translate into better outcomes. In fact, it can lead to user disengagement and even greater mistakes.

Instead, we need to simplify both the training and the tools. Here's what that looks like:

- Train with purpose, not complexity: Teach the "why" behind security, not just the "how." Show users real-world examples and consequences rather than overloading them with acronyms and protocols.
- Build secure-by-default systems: Reduce the room for human error by locking down systems by default and giving access only to what is needed—no more, no less.
- Use adaptive security models: As users grow in their roles or gain understanding, permissions and access can be expanded gradually through structured governance.
- Remove the guesswork: Interfaces and user experiences must clarify what data is being shared, with whom, and the risks.

Without thoughtful security design, AI systems will inherit the same vulnerabilities—and then scale them.

Security shouldn't feel like a test that a user can fail. It should be an invisible foundation thoughtfully integrated into the systems people rely on every day. By combining proactive training with secure design and governed flexibility, we reduce risk without burdening the people we expect to protect the enterprise.

The Analytics Wall: What's Blocking AI from Working

It bears repeating: AI thrives on clean, structured, and properly labeled data. Most organizations fall short before they even begin. According to a Deloitte survey, 62% of businesses struggle with fragmented data, making AI adoption chaotic, delayed, and often ineffective.

Fragmented data isn't just about files being in different places. It's the result of uncoordinated decisions across tools, systems, and storage. It's *how* and *where* you choose to manage your information, and more importantly—*how inconsistent that becomes overtime.*

Fragmentation comes from many sources:

- Your choice of communication or collaboration tools
- The locations where data is stored cloud vs. local, SharePoint vs. file shares
- The types of databases you use—SQL, NoSQL, third-party SaaS
- The lack of a single source of truth for document versioning
- Disconnected portals, unmanaged file-sharing platforms, or even personal cloud drives

Fragmentation isn't just a technical issue—it's a leadership one. Every disconnected tool is a decision made without alignment.

The issue isn't just technical, it's strategic. I see companies every day adding more and more tools simply because they're trendy or offer a slightly different user experience. But every new platform, every isolated decision, creates another data silo that breaks consistency, governance, and discoverability.

Here's a real example: At one organization I worked with, there were three different meeting platforms active; Zoom, Teams, and Webex. Each one stored meeting recordings, chats, and files in completely different formats and repositories. So, the question became: *How do you get all of that data into one place?* Do you build custom integrations? Manual workflows? Or worse—just ignore it?

That's not an IT strategy—it's a turf war masquerading as productivity.

And here's the harder truth: this wasn't because of a technical requirement. It was purely based on individual preferences and vendor relationships. One department liked Zoom. Another had always used Webex. IT was pushing Teams. And no one wanted to compromise.

So instead of standardizing, they fragmented.

The cost? Lost insights, duplicate meetings, unsearchable transcripts, scattered knowledge, and absolutely no foundation for AI or automation.

This is what fragmented data looks like in the real world. And its why many organizations are stuck— unable to extract value from their information, unable to govern it properly, and completely unprepared for AI.

Why Fragmentation Fails AI

Fragmentation Source	Impact on AI & Analytics
Multiple collaboration tools (Zoom, Teams, Webex)	Duplicate meetings, unsearchable transcripts
Data spread across cloud, local, and SaaS platforms	Inconsistent formats and lost metadata

Fragmentation Source	Impact on AI & Analytics
Different database types (SQL, NoSQL, 3rd-party apps)	Siloed analytics and misaligned models
No single source of truth for documents	Version confusion and broken automation
Shadow IT and unmanaged personal drives	Security risks and compliance gaps

If your teams can't agree on how to meet, store, or version documents, how can you expect AI to find or understand anything? You can't train intelligent systems on disorganized, soiled, or duplicated information. It's like teaching a GPS to navigate using ten different maps—none of which are updated.

Until leadership steps in and makes hard decisions, standardizing tools, consolidating platforms, and building toward a single data architecture— fragmentation will continue to be the enemy of progress.

Throughout this book, I've emphasized why: poor migrations, siloed environments, and scattered data sources—such as SharePoint, Google Drive, AWS, local databases, Teams, and blob storage—all contribute to enterprise confusion. When data is isolated across departments and tools, critical insights get buried, processes become redundant, and opportunities for innovation are missed.

This isn't a new problem. Think back to the era of Microsoft Access and Paradox when users would build incredibly effective databases right on their desktops. These tools often powered essential business processes, yet no one else in the company even knew they existed. These siloed solutions were never documented or shared and eventually disappeared when that employee left, or their PC was wiped. A brilliant system that got lost in the shuffle.

The issue is resurfacing now—but on a larger, more complex scale. That's why tapping into your users' knowledge is more important than ever.

Your front-line employees are the key to AI success. They understand the processes, the data inputs, and outputs, and where inefficiencies live. Think of them as embedded business analysts—people who live and breathe the workflows AI is meant to optimize. By involving them in the discovery and design phases of AI implementation, you ensure your systems are grounded, not just theory.

The takeaway: Before you rush to implement AI or analytics tools, take the time to clean up your environment, unify your data sources, and most importantly, listen to the people doing the work. Their insights can transform a struggling digital strategy into a high-performing, AI-ready enterprise.

From Chaos to Classification:
For organizations to fully leverage AI and automation, they need to:
- Clearly Define Goals – Set measurable KPIs for cost savings, efficiency, and collaboration.
- Assess Cloud Readiness – Identify gaps in current infrastructure before implementing AI.
- Align Stakeholders – IT, operations, and business leaders must work together to ensure a unified strategy.
- Invest in AI-Compatible Systems – Prioritize platforms that support automation and data intelligence.

AI is not just another tool; it's a fundamental shift in how organizations process, analyze, and act on information. Companies that ignore their potential will struggle to remain competitive.

Getting AI-ready is about structure—but staying competitive means anticipating what's next.

The Future is AI-Driven

So, what was your goal when moving to the cloud? Did your organization achieve it, or did unforeseen challenges derail the transition and vision?

The next technological shift—AI and blockchain—is already happening. To stay ahead, companies must:
- Streamline and consolidate tools
- Centralize and structure their data for AI readiness
- Adopt process integration models like PIM to automate workflows
- Rethink traditional document management and embrace AI-generated content

AI just isn't coming here. The question isn't whether legacy systems will become obsolete but how quickly companies will adapt to avoid being left behind.

Governance Maturity Model

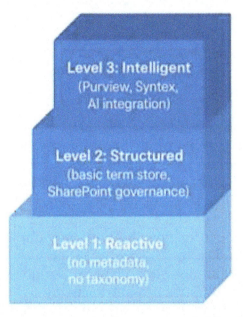

Good governance isn't about control, it's about clarity, consistency, and accountability

This means shifting mindsets—from "humans creating documents for AI to process" to "AI generating content that humans guide and approve." Knowledge work is about to become co-authored and that requires trust, governance, and structured data from the start.

If you hear that you're on the wrong path, my friend, it's time to rethink your strategy. So, you've moved everything from your on-premises file shares... Congratulations! But now you're facing a digital jungle, an environment cluttered with uncategorized, unclassified, and metadata-lacking content. It's like trying to navigate an overgrown forest without a trail map.

Tools for Transformation

As mentioned earlier, this isn't just an organizational issue, it's a matter of cost, performance, and even digital sustainability. The cloud was supposed to improve things, yet it has introduced storage sprawl, skyrocketing costs, and sluggish system performance for many companies.

Think of it this way: digital waste is the equivalent of leaving the lights on in an empty building. It drains resources unnecessarily. Excess content leads to:
- Higher storage costs in SharePoint Online
- Slower response times when searching for critical files
- Performance bottlenecks due to overloaded indexing
- Painfully slow cleanup scripts when trying to apply structure retroactively
- AI inefficiency—like giving a self-driving car the wrong GPS data

Why Cleanup Fails Without a Plan

Some organizations attempt a manual cleanup, thinking they'll sort things out over time. But let's be honest—this is entirely unrealistic when dealing with terabytes of data.

Many companies overestimate their ability to manually manage data cleanup post-migration. The reality?
- Most businesses never finish the cleanup.
- Storage bloat continues to grow.
- AI remains ineffective due to unstructured content.
- Compliance and governance risks increase.

So, if manual cleanup isn't the answer, what is?

A Strategy for Clarity: How to Clean Up & Prepare for AI

If AI works for your business, it needs structured, well-managed data. It's time to clear the overgrown digital forest and make your environment AI-ready.

Gain Visibility Over Your Content

- Implement content discovery tools to analyze what's in your SharePoint libraries.
- Use PowerShell scripts, Microsoft Purview, or third-party solutions to identify ROT (redundant, obsolete, and trivial) content.
- Audit user activity—are people still accessing old files?

Prioritize Clean Data

- Before AI can be effective, your data must be clean.
- Identify and purge unnecessary files that don't serve a business purpose.
- Archive old content instead of dragging it along in live environments.

Automate Content Classification

- Leverage AI-powered tools like Microsoft Syntex for automated metadata tagging.
- Apply sensitivity labels to ensure compliance and protect sensitive information.
- Use retention policies to delete files after a set period automatically.

Think Sustainably: Reduce Digital Waste

- Every unnecessary file stored = wasted energy and resources.
- Move away from the "just store everything" mentality.
- Treat cloud storage like a finite resource, just like energy consumption.

Bring Expertise!

- A successful post-migration cleanup isn't just about software; it requires strategy and expertise.
- If you don't have internal knowledge of structured content management, hire consultants who do.

Ready for AI? Not Until Your SharePoint, Third-Party Tools, or Data is AI-Friendly

While I often reference my experience with SharePoint and Microsoft tools throughout this book, the strategies I discuss are universal. They apply to any data platform or technology stack you use today.

My goal is not only to share my background in technology but to provide practical lessons learned so you can confidently move forward and build a real, sustainable plan for adoption of AI. With the right strategy and tools, your SharePoint Online environment can be transformed from an unstructured mess into an efficient, intelligent system that AI can leverage effectively.

But this isn't just a long-term initiative; you can start seeing results in weeks, not years.

What's at Stake?

- Lower storage costs and optimized cloud efficiency
- Faster document retrieval and indexing
- More vigorous AI-powered search and automation
- Compliance-ready data governance

Organizations that get this right will lead to AI-driven automation and digital efficiency. Those who ignore it will be stuck fighting inefficiencies, high costs, and compliance risks.

Is Power Platform Ready for Primetime?

Not yet.
The Power Platform, including Power Automate, has incredible potential for workflow automation and low-code development. However, a fundamental flaw limits its use for all business processes.

The Biggest Issue? Flow Expiration.

Power Automate flows expire after 29 days, meaning any business process extending beyond that timeframe is unsupported. Power Automate is great for short, event-driven flows. For long-running processes, Logic Apps or custom orchestration may be required.

Why This is a Problem

I've designed and reengineered numerous workflows created in SharePoint 2010 and 2013, and while SharePoint is a powerful tool, I discovered that it often falls short for many real-world business processes.

One major limitation as I stated is its handling of long-running workflows in most business processes, especially those tied to projects involving turnover or troubleshooting, extend well beyond 30 days.

This led to several recurring issues for not using Power Automate:
- Business workflows often take months, not days.
- Contract approvals, legal holds, financial audits, and extended processes will break if they rely on Power Automate.
- Organizations assume their automating processes when they're just creating new roadblocks.

231

So, What's the Solution?

If your processes require longer execution times, you need a hybrid approach using Azure Logic Apps or custom workflow solutions. Microsoft needs to rethink flow expiration limits before Power Automate can be considered a serious enterprise automation platform.

Tools for Transformation

As organizations transition to AI-powered automation, managing data integrity, security, and compliance becomes crucial. Microsoft Purview provides a robust governance framework for document lifecycle management, ensuring businesses can delete, classify, and organize content efficiently before leveraging AI. AIP focuses more on encryption and classification, whereas Purview provides holistic lifecycle governance.

Tool	Role in AI Readiness
Microsoft Syntex	Automated metadata tagging, AI-driven document processing
Microsoft Purview	Lifecycle management, retention, compliance, content classification
Power Platform	Short-term automation (with flow limits)
Azure Logic Apps	Long-running workflows and orchestration
Azure Blob Storage	Low-cost archive for non-critical or large legacy content

Why Data Cleanup Is Essential for AI

Since AI relies on structured, clean data to deliver accurate insights, automate workflows, and drive efficiency.

Many organizations rushed into cloud migration without proper cleanup, leading to:

- Excessive digital clutter – outdated, redundant, or unnecessary files.
- Unstructured content – files without metadata, making AI categorization difficult.
- Compliance risks – sensitive data not appropriately governed, leading to security vulnerabilities.

Microsoft Purview provides an essential framework to tackle these issues by enforcing legal retention policies, automated cleanups, and governance workflows.

Key Features of Microsoft Purview for Data Cleanup & AI Readiness

Advanced Retention & Disposition Policies

- Automate document lifecycle management by defining how long files should be retained before they are reviewed or deleted.
- Ensure legal compliance by meeting regulatory retention standards before data disposal.
- Align AI processes with adequately structured and legally retained documents.

Intelligent Disposition Workflows

- Microsoft Purview automates the review of files before deletion, ensuring compliance with legal and organizational policies.
- Disposition workflows act as legal checkpoints to confirm that data is reviewed and deleted securely.

Secure & Efficient Document Cleanup

AI cannot function effectively with cluttered, unstructured data. Purview enables businesses to:

- Identify redundant, obsolete, or trivial (ROT) content.
- Delete unnecessary files before AI-driven classification takes over.
- It takes 48 hours to complete tagging or deletion
- Ensure AI models are trained in relevant, high-quality data.

Governance & Compliance Through Legal Oversight

- Purview grants legal teams' direct control over document classification, retention, and deletion.
- Approval mechanisms ensure key data is correctly reviewed before deletion, reducing risk.
- Customized licensing options allow organizations to allocate governance roles efficiently.

Integration with Azure Blob Storage for Cost-Effective Governance

Not all content belongs to SharePoint or Microsoft 365. Large files (ZIPs, media, logs, etc.) can be offloaded to Azure Blob Storage, which integrates with Purview for:

- Lower-cost storage for non-critical files.
- Automated governance without overloading SharePoint storage.
- Consistent retention & deletion policies for both structured and unstructured data.

The Road to AI: Why Microsoft Purview Matters

1. Classify & organize existing data to create a structured foundation for AI automation.
2. Use Purview to clean up digital clutter before AI models analyze and automate workflows.
3. Ensure compliance by aligning AI processes with retention policies and legal governance.
4. Leverage automation to streamline document review, security labeling, and data disposition.
5. Reduce storage costs by shifting unnecessary content to Azure while keeping critical data in Microsoft 365.

Beyond Microsoft: How Other Platforms Are Adopting AI-Driven Governance

While Microsoft leads in AI-powered compliance, other cloud providers have developed similar governance tools:

- **Google Cloud AI & Data Governance**
  - Uses AI-driven classification and data lifecycle management to structure cloud-based documents.
  - Provides automated document tagging and security controls for compliance.
- **AWS AI & Compliance**
  - **AWS AI-powered data discovery** automatically labels and categorizes documents based on metadata and content.
  - Retention and disposition tools like Purview help businesses meet compliance needs.
- **IBM Watson & Enterprise Data Governance**

- Uses AI for compliance-driven automation, including contract review, legal discovery, and policy enforcement.

Why Cleaning Up Data Now Prepares You for AI

- AI needs structured, clean data—organizations that skipped proper governance are already behind.
- Microsoft Purview provides a seamless way to enforce retention, security, and compliance, ensuring that AI works with legally compliant, structured, and optimized data.
- Integrating Purview with Azure, AI-powered automation, and governance tools allows companies to scale AI while maintaining control over digital assets.

Before you deploy AI, ensure your data is clean. Purview is the key to making your AI investment successful.

Don't Let Poor Planning Derail AI

If your migration was rushed, you must fix the structure. AI cannot work without well-managed data—if your SharePoint environment is messy, AI will be ineffective. Organizations need to adopt a green mindset and cut down on digital waste.

Power Platform isn't quite enterprise-ready for long-running workflows, so alternative strategies are needed.

The time to act is now. If your post-migration environment is still chaotic, it's not too late to correct course and prepare for AI-driven efficiency.

The Road to AI Enablement

It's time to take control of your cloud environment, optimize for AI, and turn your SharePoint Online into an intelligent, efficient workspace.

The Role of SharePoint and Other Platforms in Preparing for AI

Laying the Foundation for AI Through Structured Content Management

SharePoint was pivotal in preparing businesses for AI by introducing structured content management, metadata-driven organization, and automation capabilities. While many initially viewed SharePoint as a document management and collaboration tool, it became an early precursor to AI-powered content processing.

Beyond SharePoint, significant platforms like Google, AWS, and Azure have adopted AI-driven automation and content processing, helping businesses optimize data management and efficiency. Each platform took different approaches to AI but shared the same goal: enhancing automation, searchability, and intelligent workflows.

What SharePoint and Other Platforms Brought to the Table

SharePoint's Contributions to AI Readiness
- Metadata and Content Tagging
  - Traditional file storage relied solely on folder-based organization, making search and classification difficult.
  - SharePoint introduced metadata, term stores, and content types, allowing

organizations to tag documents with structured information for better AI processing.

- Document Management & Search Optimization
  - SharePoint centralized content storage and enhanced discoverability through metadata filtering and structured search.
  - This structure became essential for AI-driven search capabilities, predictive analytics, and knowledge retrieval.
- Workflow Automation & Business Process Integration
  - SharePoint's workflows and later Power Automate helped organizations automate approval processes, task assignments, and document processing.
  - This early automation paved the way for AI-driven process optimization by reducing manual intervention and speeding up decision-making.
- Enterprise Content Management (ECM) & Compliance
  - AI relies on structured, compliant content—SharePoint introduced records management, retention policies, and auditing to support AI-enhanced compliance.
  - With tools like Microsoft Purview and SharePoint Syntex, companies can apply AI-driven classification and protection to their data.
- Preparing for AI-Driven Knowledge Management
  - SharePoint connected people to structured knowledge, helping AI models process, categorize, and analyze data for intelligent insights.

How Google, AWS, and Azure Ramped Up AI Adoption

- **Google Cloud AI & Google Drive**
  - Google took a different approach, focusing on cloud-based document storage with AI-powered search and automation.
  - Google Drive, integrated with Google Workspace AI features, allows for real-time collaboration, automatic content classification, and enhanced search capabilities.
  - Google Cloud AI provides ML-driven document processing, intelligent search, and natural language understanding, helping businesses extract insights from unstructured content.

- **Amazon Web Services (AWS) AI & Data Processing**
  - AWS focused on AI-driven automation at scale, integrating AI into cloud services, storage, and business intelligence.
  - Services like Amazon Textract allow for automatic text extraction and processing, like SharePoint Syntex.
  - Amazon Comprehend enables AI-powered text analytics, helping businesses analyze customer interactions, legal documents, and structured content.

- **Azure AI & Microsoft 365 Integration**
  - Microsoft leveraged Azure AI to enhance SharePoint's capabilities, introducing tools like Microsoft Syntex and Copilot to help businesses manage content intelligently.
  - Azure AI improves automation, predictive analytics, and cognitive search, allowing

organizations to extract insights from large volumes of data.

How SharePoint and Other Platforms Enabled AI

Metadata as the Building Block for AI

- AI needs structured data to function effectively.
- SharePoint's Metadata-driven content organization allows AI models to:
  - Understand document types and classifications.
  - Automate data categorization and tagging.
  - Improve search accuracy and relevance for AI-powered queries.

AI-Powered Search & Knowledge Extraction

- AI in Microsoft Search, Google Cloud Search, and AWS AI services use structured data to:
  - Identify relevant documents and surface them intelligently.
  - Extract meaningful insights from content repositories.
  - Enable AI-driven chatbots and virtual assistants to retrieve business knowledge.

Microsoft Syntex and AI-Enhanced Document Processing

- Microsoft Syntex, built into SharePoint, introduced AI-powered document processing and automation by:
  - Extracting key data from PDFs, invoices, and contracts.
  - Automatically tagging and classifying documents using machine learning models.

    o Apply sensitivity labels and compliance rules based on document metadata.

Comparable AI tools on other platforms:

-   o Google Document AI provides AI-powered document scanning and extraction.
    o Amazon Textract enables AI-driven OCR for structured content processing.

The Evolution of AI-Driven Content Management

- From Document Storage to AI-Driven Insights
    o Originally, SharePoint, Google Drive, and AWS S3 were designed for storage.
    o Today, they are the foundation for AI-driven document understanding, search, and process automation.

- From Manual Workflows to AI-powered automation
    o SharePoint's Power Automate, Google's AI-driven workflows, and AWS's serverless automation tools are transitioning into AI-powered process optimization.
    o AI is now predicting actions, recommending content, and improving workflows dynamically.

- From Static Content to AI-optimized knowledge
    o AI enables advanced content intelligence, allowing businesses to analyze, classify, and retrieve data at scale.
    o Platforms like Microsoft 365, Google Cloud AI, and AWS AI Services are now driving AI-powered knowledge management.

Final Thoughts: How These Platforms Are Shaping the Future of AI

- AI is only as effective as the data it processes. SharePoint, Google Drive, and AWS introduced structured data management to prepare businesses for AI.
- Microsoft Syntex, Google Document AI, and Amazon Textract are advancing AI by automating content classification and retrieval.
- SharePoint's metadata structure and Azure AI capabilities set the stage for intelligent automation and cognitive search.
- AI is no longer just an enhancement—it's an essential tool. Businesses that fail to properly clean, structure, and manage their data will struggle to leverage AI effectively.

By recognizing the foundational role of structured content management in AI, businesses can avoid the mistakes of the past and prepare for a future driven by intelligent automation. Whether through SharePoint, Google Cloud, AWS, or Azure AI, the key to AI success lies in how well data is organized, classified, and leveraged for automation.

Chapter 8:

The Road to AI and Automation

A Brief History of SharePoint

Introduced in 2001, SharePoint quickly became one of Microsoft's most successful enterprise products, evolving from a simple document management system into a robust platform for collaboration, intranet portals, enterprise content management, and process automation. At its peak, SharePoint was reportedly Microsoft's fastest-growing product ever, reaching over $2 billion in annual revenue by the early 2010s. It became the backbone of many organizations' digital workplaces, deeply embedded in enterprise IT ecosystems around the world.

Over the years, SharePoint has faced competition from various platforms trying to carve into its market share. Early contenders included IBM Lotus Notes and Documentum, which focused on collaboration and document control. Later, Google Workspace and Box provided cloud-based alternatives, while platforms like Jive, Confluence, DocuShare, FileNet, and even Slack sought to redefine team collaboration. Despite this, SharePoint retained its dominance, especially in

Microsoft-centered enterprises—due to its tight integration with Office, Windows Server, and later Microsoft 365. The evolution of SharePoint Online and the introduction of modern experiences, integration with Power Platform, and AI features like Syntex have continued to make it a cornerstone of Microsoft's enterprise cloud strategy

By 2006, collaboration in the workplace was rapidly evolving, and SharePoint was emerging as a game-changing platform. At this point in my career, I was deeply interested in how people worked together digitally. I remember coming up with the concept of collaboration through the web browser and even discussed creating a software solution that would allow users to store documents and collaborate seamlessly with tech friends. Not long after that conversation, I stumbled upon SharePoint while looking through a stack of software disks. That moment set me on a new path that would lead me to master SharePoint, build expertise in automation, and eventually start my own business.

SharePoint Evolution Timeline: From Collaboration to AI Integration

Version	Release Year	Key Milestones & Impact
SharePoint 2001	2001	First release; basic document management and web parts. Marked Microsoft's entry into collaboration platforms.
SharePoint 2003	2003	Introduced improved team sites and integration with Office. Start of enterprise adoption.

Version	Release Year	Key Milestones & Impact
SharePoint 2007 (MOSS)	2007	Major upgrade with workflows, Excel Services, BDC, and publishing features. Paved the way for enterprise content management.
SharePoint 2010	2010	Brought in the ribbon UI, sandbox solutions, and better social features. Expanded developer and admin flexibility.
SharePoint 2013	2013	Enhanced search, mobile support, and app model. Modernized user experience and cloud-ready architecture.
SharePoint 2016	2016	Bridged on-prem and cloud with hybrid capabilities. Improved stability and performance for enterprise deployment.
SharePoint 2019	2019	Final major on-premises release. Introduced modern sites, communication hubs, and deeper Office 365 ties.
Microsoft 365 +	2020s	Cloud-first, AI-integrated services like

Version	Release Year	Key Milestones & Impact
SharePoint Online		Syntex, Microsoft Viva, and Copilot. Real-time collaboration and automated governance at scale.

Diving Into SharePoint – The Early Days of Learning

Back then, there were no step-by-step tutorials, TechNet, online courses, or community forums like today. If you wanted to learn SharePoint administration, you had to figure it out yourself. So, I did precisely that.

Determined to understand the product inside and out, I went on eBay, bought four older servers, and spent two months locked in every weekend, teaching myself how to install, configure, and manage SharePoint from scratch. I researched the best practices, studied infrastructure planning, and troubleshot problems independently. It wasn't easy, but by the time I finished, I had built a firm foundation in SharePoint administration—so much so that when I went on job interviews, hiring managers would ask me for SharePoint interview questions to help them evaluate other candidates.

There were numerous integrated components within SharePoint that an administrator needed to understand, but most managers at the time didn't have a deep enough grasp of the platform to fully appreciate its complexity. Features like SharePoint Workspace, Groove, My Sites, Enterprise Search, and the Business Data Catalog (BDC) each required their own set of skills. To be an effective SharePoint administrator, you had to be proficient across a wide range of tools and technologies like Active

246

Directory and Exchange, not just the core platform. It wasn't just document storage; it was an entire ecosystem.

A SharePoint farm typically spans multiple servers, each with distinct roles—Web Front End (WFE), Search, and Application Servers. From these three foundational server types, countless configurations can be built, especially in the SharePoint 2016 and 2019 environments, where architectural flexibility increased significantly.

Governance around servers and services were more configurable in SharePoint 2016 and 2019. This gave us more control over the services that could be run on servers on our farms. Integration with other server platforms like Office Web Apps and Exchange were added bonuses but the configuration, your experience counted.

Despite the power and flexibility, SharePoint has never been an easy installation. Setting up a farm properly requires careful planning, deep knowledge of infrastructure, and strict adherence to best practices. Throughout my career, I've encountered far too many misconfigured farms, plagued with constant health alerts, performance issues, or worse, catastrophic failures. Some of those issues were quick fixes; others were far more damaging.

One of the worst cases I saw involved an administrator who installed the entire SharePoint farm using his personal Active Directory user account. What he didn't realize is that SharePoint setup requires a dedicated installation and service account for best practice. When that administrator eventually left the company, his account was disabled by the AD team. As a result, the entire SharePoint farm went down. When they reached out for support, we identified the root cause and had to request a temporary reactivation of the user account just to bring the environment back online.

The takeaway? Know your installation procedures, understand the platform's best practices, and always think long-term when configuring enterprise systems. A shortcut today can cost the business dearly tomorrow.

Legacy vs. Modern Tools: SharePoint & Enterprise Tech Evolution

Legacy Tool/Method	Modern Equivalent	Use Case / Improvement
InfoPath	PowerApps	Custom forms with responsive design, mobile support, and integration with Dataverse.
SharePoint Designer	Power Automate	Modern, cloud-based workflow automation and event-driven triggers.
FTP (Port 21)	SFTP / Azure Blob Storage	Secure file transfers with encryption and cloud-native integration.
Folders	Metadata / Managed Metadata Service	Smarter search, filtering, and classification—critical for AI-readiness.

Legacy Tool/Method	Modern Equivalent	Use Case / Improvement
Email Approvals	Adaptive Cards / Teams Approvals	Streamlined, trackable approvals integrated into collaboration platforms.
Custom Web Parts (Sandboxed)	SPFx (SharePoint Framework)	Modern, scalable development with React and better governance.
Excel Macros	Power BI / Power Query	Visual analytics and advanced data modeling in real-time dashboards.
Hardcoded Permissions	Azure AD Groups / Sensitivity Labels	Centralized, conditional access and policy-based security.
On-Prem File Shares	SharePoint Online / OneDrive	Anywhere access, version control, co-authoring, and integrated governance.
STSADM / Manual Scripts	PowerShell + PnP + CLI for Microsoft 365	Scripting and automation with community-backed tools and API access.

Insight: Modern tools aren't just replacements, they're foundational building blocks for scalable, secure, and AI-ready enterprises.

The SharePoint Interview That Proved My Expertise

One interview stood out. The hiring manager told me about a significant issue their team had struggled with for weeks. They even called Microsoft Support, and after multiple attempts, Microsoft couldn't fix it.

As soon as he described the problem, I smiled because I had solved the same issue for another client just a week prior. After I relayed this to the IT Manager, he called the SharePoint Administrator to the office. I told the SharePoint Administrator precisely what to do in the interview and it worked.

The manager was stunned, but I didn't get the job despite this. The other executive in charge had his own agenda and focused on project management topics, which had nothing to do with the SharePoint role I was applying for. It had nothing to do with my resume as well. It was clear that the hiring manager wanted me for the job, but the big boss had other ideas.

Still, that experience validated just how skilled I had become in SharePoint. You know you've reached a new level when you can walk into an interview and fix an issue the company's IT team had for over two months and Microsoft couldn't resolve.

Mastering Challenges: A Journey Through SharePoint's Evolution

As a consultant, I worked with MOSS 2007 (Microsoft Office SharePoint Server), a transformative period for enterprise collaboration. SharePoint was gaining traction across industries, and I was at the center of its adoption.

One of my most memorable projects was in 2009 at a customer in Las Vegas, where I was hired as a SharePoint consultant for a 6-month period. It was a fast-paced role, with the company eager to move away from traditional file shares and implement a more structured collaboration system.

I remember one instance when we had just rolled out a new workflow in SharePoint 2007 MOSS. We had tested it thoroughly for two weeks, and everything worked as expected. But the moment we launched the site, the workflow stopped working.

Back in those early SharePoint days, it wasn't uncommon for patches to break features, and this was one of those moments. After digging into the servers, we discovered that an update had automatically installed the night before. Sure enough, Microsoft had already posted an alert on their site about the issue, along with a note that a fix would be published later that day.

It doesn't happen as frequently now, but in those days, these kinds of disruptions were common. You had to stay sharp, think on your feet, and always be ready with a workaround. SharePoint kept you on your toes—and taught you how to pivot fast when systems didn't behave the way they did in testing.

After my time there, I transitioned to a project in South Carolina, where I developed an entire intranet and extranet for the entity. The project was highly competitive—there were 60 applicants for the role, and I landed the job. This was a pivotal moment for me, proving that my expertise in SharePoint, workflows, and governance set me apart.

While I worked with many tools during this time, SharePoint was always at the core. I became a master of PowerShell and CAML queries. As I learned more, I became more confident in architecting solutions and helping organizations move beyond essential document storage to real workflow-driven environments.

Getting the Call from The Tech Leader

While in South Carolina at a customer site, I received a call from a leading tech company for a job interview. I had been working toward this opportunity for years and was determined to get the role.

During the interview process, I learned a valuable lesson—one that many people in tech still struggle with today:

If you don't know something, say you don't know.

At one point, I was asked a technical question about SharePoint searching, something I had not yet worked with at a deep level. Instead of trying to fake my way through it, I said, "I don't know, but I would love the opportunity to learn." That honesty earned the respect of my interviewer, and I moved forward to the final interview round.

I ended up getting the job, and it was an incredible experience. Working for a tech leader meant I would travel, work with enterprise clients, and help organizations implement and optimize SharePoint on a massive scale.

The SharePoint Boom: Seeing the Industry from Inside a Tech Leading Company

While traveling with a Tech Leading Company, I saw every kind of SharePoint deployment imaginable, from well-planned, highly successful implementations to complete disasters.

Some deployments failed due to misconfigurations—small mistakes in setup that cascaded into significant issues. Others failed due to overextension companies trying to make SharePoint do things it wasn't designed to handle. Some failures were due to user resistance—companies would launch a new SharePoint site, but employees wouldn't use it because they weren't appropriately trained or didn't see its value.

And then there were the administrators; some were highly skilled and well-prepared, but many had no real experience with the platform and took on several roles, including SharePoint.

We saw a trend in which people who knew only a few SharePoint buzzwords were getting jobs as "SharePoint Admins." Companies lacked the expertise to vet candidates properly, so they hired people who barely knew how to configure a basic SharePoint site.

This same problem is happening now with AI—people with no fundamental technology background are branding themselves as AI experts because it's a new field or because their boss wants them to be.

If you are hiring for AI, do not bring in someone without at least a few years of technology experience. If you do, you'll end up with poorly designed solutions that will create significant problems in the future.

The Pressure to Implement Fast – A Recipe for Failure

While working at the leading company, I quickly realized that most SharePoint failures weren't technical failures but leadership failures.

- Executives wanted SharePoint "now."
- They didn't wait. They didn't plan.
- They didn't follow the best practices.
- They hired someone who managed multiple applications.

Instead of ensuring a proper rollout, many organizations rushed through implementations, leading to the following:

- Unstructured sites with no governance.
- Admins with no training.
- Users who didn't understand the system and didn't use it.

This problem still exists today.

Managers pressure teams to roll out systems too quickly, leading to disastrous implementations, rushed migrations, and frustrated employees. They assumed that just because the system was installed, everyone would magically know how to use it. However, adoption was low without training, and the investment was wasted.

This experience taught me that successful technological adoption isn't just about the tool it's about the people. If you don't educate your users and give them the proper guidance, the best technology in the world won't help.

The Lessons I Took from the Leading Tech Company

My time at the Leading Tech Company solidified everything I had learned over the years:

- Honesty is critical in tech – If you don't know something, admit it, then go learn it.
- Hiring matters – Bringing in underqualified people leads to massive failures.
- Planning beats speed – Rushing into technology adoption without governance leads to chaos.
- Training is essential – No system succeeds if people don't understand how to use it.

The leading tech company was one of the best experiences of my career. It exposed me to the highest levels of enterprise IT and showed me that technology success is as much about strategy and leadership as it is about coding and configuration.

I wasn't chained to a desk with a boss hovering over my shoulder. The company trusted me to manage my time, engage with customers, and deliver results. It was a refreshing experience, proof that not every organization needs to micromanage. Sometimes, you just need to hand over the keys and let people drive.

I had a real passion for the work I was doing and genuinely wanted to help that company succeed. I talked about it everywhere, on planes, at parties, in passing conversations. I even convinced people who once dismissed SharePoint to give it a second look, simply by connecting it to the real problems they were facing in their organizations.

One conversation that stands out happened on a flight, where I sat next to a CEO who wasn't sold on SharePoint. We ended up talking for the entire two-hour flight and by the end of it, he was ready to take another look. That's the kind of impact passion and trust can have.

From this point forward, I knew that automation, AI, and governance would be the future—and I was determined to be part of that transformation.

Building My Own Business and Writing the Book

That next phase of my journey came with a defining shift: I launched my own SharePoint consulting business.

Prior to working at the Leading Tech Company, I worked at a few companies from various industries. I had a brief stop at an identity management company where I learned about smart cards for the first time. It really opened my eyes to the new wave of security coming.

I then landed a contract role as a SharePoint Subject Matter Expert (SME) at an agency in 2006. It was a pivotal moment. Not just because of the role itself, but because it showed me, I could work independently, manage client expectations, and build a business around my own knowledge. I wasn't just supporting systems anymore, I was guiding strategy, leading implementations, and proving that my experience could stand on its own.

That season cemented my passion for collaboration, automation, and enterprise systems. SharePoint, to me, was never just a document repository. It was a framework for structured collaboration, governed workflows, and process automation, all things that would later lay the groundwork for AI-driven enterprise systems.

After working at the Leading Tech Company and before I ever wrote a book or started working independently, I worked on contracts across a wide range of industries from federal and state government agencies to global enterprises and small businesses from 2013 - present. That exposure gave me a well-rounded view of what organizations are really facing.

I kept detailed notes from every engagement going back to 2009. And when I look at them now, it's almost eerie how many of the same challenges are still present today: rushed implementations, reactive decisions, lack of planning, no clear requirements, and pressure to "just get it done." The tools may have changed—but the patterns haven't.

I took on projects that pushed me beyond my comfort zone, even when I wasn't fully familiar with the technology at hand. I welcomed challenges not because I had all the answers, but because I wanted to learn. And I did. I delivered it repeatedly. I overcame obstacles. I built systems, trained users, solved infrastructure issues, and navigated team dynamics all while continuing to grow.

I remember taking on a major challenge for an educational company in Maryland that was trying to leverage Business Data Catalog (BDC), a powerful but complex feature in SharePoint 2007 MOSS. At the time, resources were scarce. There was no YouTube, and there weren't many blogs or community posts detailing real-world implementations of BDC, which made it even more daunting.

But I saw it as an opportunity, and I was not scared to stretch my knowledge and push my skills further. I took the project head-on, worked through the bumps, and did extensive testing in a development environment. In the end, the effort paid off. I was able to successfully implement the solution, fully utilizing BDC to connect and surface external business data right within SharePoint.

That willingness to take on challenges started long before my tech career. One of the earliest moments I can remember was back in my football days.

We were preparing for a Friday night game against an undefeated team with a running back who was supposedly the fastest in the league. All week, that's all we heard about. Our coaches couldn't stop talking about how good he was, how dangerous, how untouchable. I don't know if they were trying to motivate us or if they truly believed we didn't stand a chance, but their energy wasn't exactly confident. Practice felt flat. We were being set up to expect defeat.

Game night came, and early on, my coach called a play down and out for me. I got to the line and looked up to see none other than *that* guy across from me. The one we'd been told to fear. The league's fastest. I didn't flinch. I didn't back down. I simply said to myself, *"I'm going to beat him."*

The ball snapped. I ran my route. Caught the pass at the 24-yard line. And then I kept going. Seventy-six yards later, I was in the end zone and that so-called fastest guy. He never caught me.

I'm not sharing that to brag. I share it because the lesson stuck with me: challenges will always come, and it's not about whether you feel ready, it's about whether you believe you can rise to the moment. That same mindset has carried me through years of IT work. In tech, just like on the field, you can either shrink back or show up. I chose to show up and keep running.

That drive and being a musician in that unstable environment is what gave me the courage to go independently.

During this period, while starting my own business in 2013, I consulted with nearly 100 companies worldwide at this point in my career due to working at the leading tech company. I traveled extensively, training teams, advising on architecture and governance, building solutions, and stepping into situations that others might have avoided. Whether the issue was collaboration breakdown, system sprawl, or bad leadership, I chose to be on-site, to show up in person, to learn what I didn't know and share what I did.

One moment I'll never forget was during a remote project where I was tasked with building a new intranet page for a major company. On a call with over 20 people, including leadership, the CTO called me a liar.

It shocked me.

She always was very abusive on the phone to everyone on the calls. In this instance she claimed she had given me specific instructions, which she hadn't. There were no written requirements, no documentation, no clear direction. The project was being run from the top of her head. And when things didn't align with her vision, the easiest thing to do was shift blame.

I hung up the call because she was very angry, she usually was on every call. Maybe not the most diplomatic move by me but after that I was communicated to through another resource to give me message from the CTO. I still finished the project with flying colors and handed off a beautiful Intranet site in SharePoint.

That moment wasn't just about disagreement, it was a wake-up call about how organizations treat timelines, documentation, planning, leadership, and collaboration. I've worked with many CTOs, CEOs, and C-suite leaders over the years, and a pattern emerges: hitting the deadline matters more than doing the work right. There's no room for planning, documentation, design changes, infrastructure limitations, or even better ideas. The only goal is to be "on time."

That mindset is broken.

We cannot expect SMEs and technical experts to be mind readers. We cannot build collaboration while allowing toxic leadership or inflexible schedules to poison the process. And we cannot build sustainable systems without slowing down to get it right.

These experiences shaped how I think about leadership, communication, and design today. They also deeply influenced what we're building with Tourque— our AI Operating System for organizations. We didn't just design Tourque to automate tasks. We designed it to protect collaboration, support technical clarity, and reduce the friction that comes from poor management decisions.

AI will do more than anyone expects in the next phase of technology.

My solution Tourque is being built to lead the way

In 2020, I was given the opportunity to write a book:

Explained Simply. Designed for Real Users. Built for Results

It was exciting. It was also one of the most challenging projects I've ever taken on.

There were many moments I wanted to quit, tight deadlines, constant pressure from the publishing team, and the fear of making a mistake in print. But I kept going. Because I believed in the message behind the book.

I didn't want to write a deeply technical manual. I wanted to make SharePoint approachable for people who weren't tech-savvy, who just needed someone to explain it in plain language. That's always been one of the biggest gaps I've seen in the field: technical professionals who can build solutions, but struggle to explain them in terms others can understand.

The book offered practical guidance—real-world use cases, introductions to SPFx web parts, Power Platform integration, and emerging tools that were just gaining mainstream traction. Although it was released in December 2020—after SharePoint 2019 had already started to phase out, it still found its place with those seeking on-premises collaboration solutions.

The book didn't become a bestseller. I did the best with the timeline I had, and it did leave out time to review but it changed the way I saw SharePoint—and the industry.

I realized we were entering the end of the on-prem era. And with that, I saw SharePoint not just as a content management system—but as my personal bridge to the future. It became the platform that launched my journey into automation, enterprise architecture, and ultimately, artificial intelligence.

What is AI? Understanding the Core Components and How to Leverage Them for Success

Artificial Intelligence (AI) is not just one thing; it's a combination of technologies that allow computers to simulate human intelligence and make data-driven decisions. AI has become integral to business strategy, automation, and digital transformation. However, AI is only as effective as the data it is fed.

To fully understand AI and how to leverage it effectively, we must break it down into its core components, data requirements, business applications, and the tools available to help organizations optimize their AI journey.

The Accidental AI Admin – Risks of Inexperience

What is an "Accidental AI Admin"? A person suddenly tasked with deploying AI tools in the enterprise, despite having little to no experience with data architecture, governance, or structured collaboration systems.

Symptom	Impact
Deploys AI without preparing the data	Poor results, hallucinations, and inconsistent outputs
Doesn't understand metadata or taxonomy	AI can't classify, route, or search documents effectively
Confuses automation with intelligence	Build robotic workflows, not smart decision-making engines
No background in security or compliance	Exposes the org to legal, regulatory, and reputational risk

Lesson: If someone isn't ready to manage SharePoint governance, they are not ready to manage AI.

The Core Components of AI

AI is a broad term encompassing multiple computer science disciplines and data processing. The key components include:

- Machine Learning (ML) – AI's ability to learn patterns and make predictions based on data.
- Natural Language Processing (NLP) – AI's ability to understand and generate human language.
- Computer Vision – AI's ability to analyze and interpret images and videos.
- Robotic Process Automation (RPA) – Automating repetitive tasks using AI-driven workflows.
- Neural Networks & Deep Learning – Complex algorithms that mimic the human brain to process large amounts of data.

Each component is crucial in enabling AI-driven automation, decision-making, and problem-solving. However, without clean and structured data, AI cannot function effectively.

The Power of Storytelling and Sentiment: Human Connection Meets Machine Insight

Why Storytelling Still Wins

Storytelling isn't just for branding, it's a strategic tool for clarity, alignment, and emotional connection. In business, a well-crafted narrative helps customers, employees, and stakeholders understand why you do what you do, what you stand for, and where you're going.

In the age of AI, storytelling becomes even more powerful. When businesses embed stories into their digital workflows—goals, values, competitor awareness, they give AI something to learn from. AI can then use those stories to shape recommendations, anticipate needs, and align future content with your strategic vision.

How Storytelling Supports AI-Driven Strategy

When you articulate your company's mission, challenges, and customer success stories clearly, you give AI a framework to build on.

This storytelling foundation can power:

- **Competitive analysis**: Feed AI models examples of customer wins, competitive positioning, and lost deals to train them in what works.
- **Content generation**: Story-based prompts improve the quality of AI-generated messaging, proposals, and training materials.
- **User alignment**: Align internal communication with business goals by embedding stories onboarding, team updates, and documentation.

Sentiment as a Business Signal

In parallel with storytelling, sentiment analysis offers immediate, measurable insight into how your audience is feeling, not just what they're saying. It's an emotional thermometer for your digital presence.

During one project, I experimented with a Power Automate flow that monitored Facebook comments on our company's posts. Using AI-powered sentiment analysis, the system would detect whether a comment had a positive, neutral, or negative tone. If the sentiment was negative, it triggered an automatic email alert—so we could respond immediately, defuse tension, or show the customer we were listening.

These kinds of tools are game changers. They allow businesses to be more responsive, human, and intentional, at scale.

AI-Powered Sentiment Tools Can:

- Flag customer dissatisfaction before it escalates
- Track emotional trends over time to guide messaging and campaigns
- Score feedback from surveys or reviews to improve service models
- Align your internal tone with external brand promises

Key Takeaway:

Storytelling provides context. Sentiment adds emotional intelligence. Together, they enable a responsive and engaging business environment.

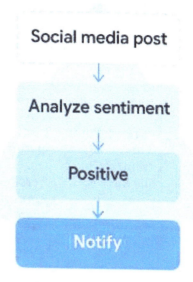

Power Automate Sentiment Analysis

The Role of Clean Data in AI Success

I know I've emphasized this throughout the book—and you'll continue to see it in every chapter on AI—but it bears repeating clean, structured data is the most critical foundation for successful AI integration. AI is only as bright as the information you give it.

If your system doesn't know what a document is or how it fits into your environment, it will never understand how to use it effectively. That's why organizing and categorizing your data isn't optional; it's critical.

Take the time to locate files and datasets, bring them together, and apply consistent structure and labeling. This step alone will make or break your AI journey.

Collaboration-to-AI Maturity Model

Stage	Name	Tools/Technologies	Key Capabilities	Common Pitfalls
1	Collaboration Maturity	SharePoint (On-Prem/Online), File Shares	Document storage, permissions, version control, team sites	Folder sprawl, no metadata, poor governance
2	Workflow Automation	Power Automate, Nintex, SharePoint Designer	Automated approvals, notifications, form submissions, task routing	Unmaintainable flows, reliance on one power user
3	AI Assistance & Optimization	Copilot, Microsoft Syntex, Purview, Azure AI, OpenAI API	Content summarization, intelligent metadata tagging, compliance automation, predictions	AI built on messy data, poor integration strategy

Pro Tip: *You cannot skip steps. AI will not work on top of chaos. Structure comes before automation, and governance comes before intelligence.*

Is AI Effective Without Clean Data?

No. AI relies on structured and well-labeled data. AI will struggle to process and analyze information efficiently without proper data governance, metadata, and organization.

How Does Clean Data Improve AI?

- Higher Accuracy – AI can generate better insights when it processes clean, structured, and well-organized data.
- Better Automation – AI-powered workflows function seamlessly when the underlying data is appropriately classified and managed.
- More Effective Search and Analysis—Optimized data makes AI-driven search engines, content processing, and analytics tools more useful.

Key Steps to Preparing Data for AI

- Organize & Structure Your Data – Implement metadata, taxonomies, and governance policies.
- Eliminate Redundant, Obsolete, and Trivial (ROT) Data by cleaning up cluttered SharePoint sites, file shares, and databases.
- Use AI-Driven Classification – Tools like Microsoft Syntex and Microsoft Purview can help tag and classify documents for better AI integration.
- Ensure Compliance and Security – Protect sensitive data and enforce security policies before AI adoption.

How Can AI Be Used? Practical Applications

AI is already revolutionizing business operations across industries. Here are some practical ways companies are leveraging AI today:

Business Automation & Workflow Management

- Automated document approvals using Microsoft Power Automate & AI-powered workflows.
- Email sorting & response automation to improve efficiency.
- Intelligent contract management for reviewing legal documents.

Data Analytics & Insights

- AI-driven business intelligence using Microsoft Power BI, Google Analytics, and Tableau.
- Predictive analytics to anticipate trends and market shifts.
- Customer behavior analysis for marketing strategies.

Content & Document Management

- AI-powered search capabilities in Microsoft 365 & SharePoint Online.
- Automated metadata tagging for content categorization and search optimization.
- Transcription & translation services to improve accessibility.

Security & Compliance

- Fraud detection in financial transactions.
- AI-powered risk assessments for cybersecurity.

- Automated data classification & labeling using Microsoft Purview.

What AI Tools Are Available?

Many AI-powered tools are available today, designed to streamline workflows, improve productivity, and enhance security. Some of the most notable AI platforms include:

Microsoft AI & Cloud Tools

- Microsoft Syntex – Automates content classification and metadata tagging.
- Microsoft Purview – AI-driven data governance, compliance, and security.
- Azure AI Services – Custom AI models for vision, language, and machine learning.
- Power Automate – AI-enhanced workflow automation.
- Copilot in Microsoft 365 – An AI-powered productivity assistant for Office apps.

AI for Search & Knowledge Management

- ChatGPT & OpenAI API – AI-powered conversational assistants.
- Elasticsearch & AI-driven search engines for enterprise solutions.

AI for Cybersecurity & Compliance

- Microsoft Defender & Sentinel – AI-driven threat detection and response.
- Google AI & IBM Watson AI – AI for data security & business intelligence.

Final Thoughts: Preparing for AI-Driven Business Transformation

AI is not just the future; it's happening now. However, businesses that rush into AI adoption without preparing their data and infrastructure will struggle to see meaningful results.

- Ensure your data is structured and clean before implementing AI.
- Choose AI tools that align with your business needs and integrate with your existing platforms.
- Train your teams on AI-driven workflows to ensure adoption and efficiency.

AI is not magic; it requires strategy. By taking a structured approach, companies can unlock AI's full potential, making their operations more innovative, efficient, and future-ready.

Collaboration as the Bridge to AI & Automation

In the modern digital landscape, collaboration has always been a critical component of technological advancement. From the early days of document sharing in SharePoint to the rise of real-time co-authoring and AI-driven automation, collaboration tools have set the stage for AI integration. However, many organizations fail to recognize collaboration as the first step in AI readiness, leading to fragmented implementations and inefficiencies.

To fully harness AI and automation, businesses must master collaboration—ensuring that data, workflows, and communication are structured, accessible, and well-integrated.

How Collaboration Paved the Way for AI

From File Sharing to Intelligent Workflows

- Early collaboration was essential for file sharing without metadata, versioning, or automation.
- SharePoint introduced a structured collaboration with document libraries, metadata tagging, and version control, preparing businesses for AI-driven content management.
- Today, AI tools like Microsoft Syntex, Google AI, and AWS AI extract insights from structured data, proving that early collaboration strategies shaped AI's ability to process information intelligently.

Metadata and Content Structuring: The Foundation of AI

- AI requires structured data to function effectively, and early collaboration tools introduced metadata-driven organization.
- SharePoint's term store, managed Metadata, and taxonomy services helped businesses categorize content, later enabling AI models to:
  - Predict document classifications based on historical patterns.
  - Improve search accuracy using structured metadata.
  - Automate compliance processes through AI-powered labeling.
- Other platforms like Google Drive and AWS AI followed suit, using machine learning models to categorize and tag documents automatically.

AI-Powered Collaboration: The Next Evolution

- Microsoft, Google, and Amazon have integrated AI-powered collaboration tools into their ecosystems:
  - Microsoft Copilot in SharePoint, Teams, and Word provides AI-generated summaries and content recommendations.
  - Google Workspace AI suggests responses in Gmail, auto-generates documents and enhances video meetings.
  - AWS AI-powered automation allows for intelligent routing, workflow optimization, and content recognition.
- These AI-driven tools enhance collaboration by reducing manual tasks, streamlining workflows, and enabling real-time decision-making.

Roadblocks: Why Many Companies Struggle to Implement AI Through Collaboration

Even though collaboration set the foundation for AI, many businesses failed to follow the best practices, leading to challenges in AI adoption:

Lack of Structured Content

- Many organizations still use traditional folder-based storage instead of metadata-driven collaboration.
- This creates a disorganized digital workplace, making it difficult for AI to process data effectively.

Rush to Migrate Without Strategy

- Cloud migrations dump all content into SharePoint Online, Google Drive, or AWS storage without curating or organizing.
- AI performs best with clean, structured data, yet many businesses migrated without governance, leading to inefficiencies.

Failure to Integrate AI with Collaboration Tools

- Businesses often view AI as an add-on rather than an integrated part of their collaboration ecosystem.
- Without a seamless integration strategy, AI tools remain underutilized or ineffective.

How to Optimize Collaboration for AI & Automation

For AI to truly enhance collaboration, businesses must revamp their approach to digital workspaces:

Implement AI-Ready Collaboration Tools

- Microsoft SharePoint & Teams + AI (Copilot & Syntex)
  - Use AI-powered search, metadata automation, and intelligent workflows to enhance collaboration.
- Google Workspace AI & AWS AI
  - Leverage auto-generated content, AI-driven document processing, and smart search recommendations.

Adopt Metadata & Taxonomy Best Practices

- Implement structured metadata and content classification to help AI understand and retrieve data efficiently.
- Use content-type hubs, term stores, and AI-assisted tagging to streamline collaboration.

Automate Workflows for AI-Driven Efficiency

- Use Microsoft Power Automate, Google App Script, or AWS AI workflows to automate document approvals, content categorization, and notifications.
- AI should support collaboration by minimizing manual intervention and streamlining decision-making.

The Future: AI-First Collaboration & Automation

AI-Driven Collaboration Workspaces

- Future collaboration tools will be AI-first, integrating predictive analytics, content summarization, and workflow automation.
- AI will eliminate redundant meetings by analyzing conversations, extracting key points, and auto-generating action items.

Seamless AI & Automation in Business Processes

- AI will not just suggest action it will execute them.
- Instead of waiting for approvals, AI will automatically analyze patterns and approve common requests.

- AI-driven workflows will enhance efficiency by routing tasks intelligently across teams.

AI Will Adapt to User Behavior & Preferences

- AI will learn from user interactions and personalize collaboration experiences.
- Smart content assistants will predict what information users need before searching for it.

Collaboration as the Gateway to AI

- Collaboration was the first step toward AI adoption—businesses that structured their digital environments correctly are now ahead of the curve.
- Platforms like Microsoft SharePoint, Google Drive, and AWS AI are evolving from collaboration tools to AI-driven knowledge hubs.
- To succeed in AI & automation, businesses must fine-tune their collaboration strategies, adopt structured content practices, and integrate AI seamlessly into their workflows.

By bridging collaboration and AI, companies can unlock the full potential of automation, efficiency, and intelligent decision-making, setting the stage for the next digital transformation era.

This chapter should showcase how collaboration tools like SharePoint transformed the enterprise landscape, leading to the AI-driven workplace we see today. Microsoft made big wins and mistakes, but SharePoint's dominance proved that structured collaboration is essential for automation and AI-driven business processes.

The Hidden Challenges Enterprises Face in AI Transformation

As enterprises race to implement AI, they often overlook critical challenges that can derail their transformation efforts. From employee adoption to governance and security risks, these hidden obstacles can create inefficiencies, increase costs, and expose organizations to cyber threats. Understanding and proactively addressing these challenges is essential for achieving long-term AI success.

Employee Adoption: Overcoming Resistance to Change

The Problem

As we saw in my previous chapters, employees frequently resist new technologies for fear of redundancy, lack of training, or unclear guidance. The introduction of AI and automation is no exception. A McKinsey study found that 70% of digital transformation efforts fail due to employee resistance and inadequate change management. Many organizations assume employees will naturally adapt, but adoption rates remain low without structured support.

Why It's Overlooked

- Leaders prioritize technology adoption over workforce engagement.
- Employees revert to old workflows when they don't see the value in new tools.
- Organizations fail to address **job security concerns**, leading to quiet resistance.

Solutions

- Invest in comprehensive training programs to ensure smooth adoption.
- Involve employees early in decision-making to foster engagement.
- Ensure new tools solve long-term problems rather than acting as short-term fixes.

Case Study: A global retail company implemented an AI-powered CRM but failed to train employees properly. As a result, workers reverted to using spreadsheets, leading to a 30% adoption rate and millions in wasted investment.

Data Clean-Up: The Foundation for AI Success

The Problem

Unstructured, bloated data slows down AI initiatives. Many organizations migrated to the cloud without cleaning up their legacy data, creating inefficiencies and inaccuracies that AI cannot effectively process. Gartner estimates that poor data quality costs businesses an average of $12.9 million annually.

Why It's Overlooked

- Data clean-up is tedious and often deprioritized.
- Organizations assume AI can "fix" messy data instead of preparing structured, accurate inputs.

Solutions

- Implement AI-powered automation to detect and categorize outdated files.
- Adopt systematic data governance policies to maintain clean, accurate records.
- Designate data stewardship teams to ensure ongoing data hygiene.

Lesson Learned: Garbage In, Garbage Out! Without structured data, AI cannot deliver accurate insights.

Enterprise Systems Complexity: The Integration Nightmare

The Problem

Overlapping, redundant, and outdated enterprise systems complicate AI adoption. Companies often patch legacy systems instead of investing in scalable, cloud-first platforms, leading to high maintenance costs and integration failures.

Why It's Overlooked

- IT teams prioritize new features over system simplification, worsening the complexity.
- Organizations delay modernization due to fear of disrupting business operations.

Solutions

- Consolidate redundant systems to streamline operations.
- Eliminate obsolete platforms that do not support AI and automation.
- Prioritize integration-ready solutions to enable smooth AI adoption.

Case Study: A financial services company relied on seven overlapping CRMs, creating data inconsistencies and AI inefficiencies. After consolidation, productivity increased by 35%, and maintenance costs dropped by 50%.

Security Risks in the AI Era: New Threats, Old Defenses

The Problem

Legacy cybersecurity protocols are not designed to defend against AI-powered attacks. As hackers leverage AI for automated cyberattacks, outdated security measures expose enterprises to high-speed, large-scale threats.

- Are we still using legacy SSL or FTP?
- Are our AI models being audited?
- Do we have blockchain-based integrity validation?

You could use the Darktrace example to introduce a broader point: "AI fights fire with fire—only AI can detect AI-powered threats in real-time.

Why It's Overlooked

- Many companies deprioritize security until a breach occurs.
- AI-powered cyberattacks can exploit legacy vulnerabilities in seconds.

Solutions

- Adopt AI-driven security tools that proactively detect and mitigate threats.
- Strengthen Identity and Access Management (IAM) to prevent unauthorized access.
- Regularly update security infrastructure to eliminate vulnerabilities before they are exploited.

Case Study: A financial institution suffered a $50M data breach due to legacy security flaws, causing reputational and regulatory damage.

Governance and Compliance Gaps: The Silent Risk

The Problem

Many organizations underestimate the complexity of global compliance laws, leading to reactive rather than proactive governance strategies.

Why It's Overlooked

- Governance is often an afterthought, addressed only when compliance audits arise.
- AI introduces new compliance risks, including bias, data privacy, and automated decision-making concerns.

Solutions

- Establish enterprise-wide governance policies from the start.
- Use AI-driven compliance tools to automate audits and policy enforcement.
- Educate employees in compliance with best practices to prevent inadvertent violations.

Key Stat: Regulatory fines for data breaches have increased by 45% since 2021, with GDPR alone issuing $4 billion in fines over the past three years.

Cloud Migration Costs: The Hidden Expenses

The Problem

Many enterprises underestimate the cost of cloud migration, leading to budget overruns from data transfer fees, misconfigured instances, and storage bloat.

Why It's Overlooked

- Organizations focus only on subscription costs, ignoring long-term operational expenses.
- Poor data hygiene leads to wasted storage and excessive egress fees.

Solutions

- Conduct a Total Cost of Ownership (TCO) analysis before migrating.
- Optimize cloud storage usage to reduce redundant data.
- Use cost monitoring tools like AWS Cost Explorer, Azure Cost Management, or Google Cloud Billing.

Case Study: A healthcare provider tripled their expected cloud costs after failing to clean up redundant files before migration.

Collaboration Breakdown: Silos and Communication Failures

The Problem

Poor collaboration between departments leads to inefficiencies, duplicated work, and fragmented AI initiatives. Even with advanced collaboration tools, silos remain a significant roadblock.

Why It's Overlooked

- Organizations prioritize system upgrades over fostering a culture of collaboration.
- Employees continue using outdated methods, resisting cross-functional integration.

Solutions

- Implement cross-departmental collaboration tools to unify teams.
- Encourage leadership to actively support knowledge-sharing initiatives.
- Foster a culture where employees embrace AI-enhanced collaboration rather than resist change.

Key Stat: 70% of employees feel left out of decision-making processes despite having collaboration tools available (Gartner).

Moving Forward with a Sustainable AI Strategy

The biggest mistake enterprises make when adopting AI is focusing solely on technology while neglecting people, processes, and governance.

Key Takeaways for a Sustainable AI Strategy:

- Recognize hidden barriers before they become major roadblocks.
- Invest in structured employee training and upskilling programs.
- Prioritize data cleanliness and governance to enable AI-driven transformation.
- Adopt AI-powered security solutions to mitigate next-gen cyber threats.
- Ensure cross-departmental collaboration and knowledge-sharing.

Final Thought: The enterprises that successfully navigate these hidden challenges will be best positioned for the AI era. They will leverage efficiency, security, and scalability to outperform competitors and drive sustained innovation.

Additional Hidden Challenges to Consider

- Shadow IT: Unregulated tools introduce security risks.
- Integration Fatigue: Overuse of third-party integrations slows down innovation.
- Vendor Lock-In: Over-reliance on single vendors limits flexibility.
- AI Misuse: Poorly implemented AI leads to bias, inefficiencies, and compliance risks.
- Legacy Mindsets: Resistance to change stalls progress.
- Time-to-Value Misalignment: Unrealistic ROI expectations hinder adoption success.

By preparing for these challenges, enterprises can ensure a smoother, more effective transition into the AI-driven future.

Why Isolated AI Solutions Can Kill Your Business: The Hidden Danger of Siloed AI Applications

We've seen this pattern before. A surge of new apps hits the market—each promising innovation, speed, and simplicity. Companies rush to adopt them, often without strategy or governance, and before long, the organization becomes fractured. Teams begin working in silos, using disconnected tools like Asana, Teams, or Slack with no centralized communication or oversight. I've been on the ground with organizations where it was a complete free-for-all—no standards, no integration, just chaos. Productivity suffered, and the alignment across departments collapsed.

As AI embeds itself into every tool, the risk is even greater. Isolated AI apps might offer short-term gains, but without integration into your enterprise ecosystem, they become distractions and liabilities at worst. If different departments deploy their AI tools without oversight, the result will be fragmented data, duplicate efforts, and missed opportunities.

Take control now. Establish strict guidelines on which tools and platforms are authorized. Don't repeat the mistakes I've seen time and again. Companies that win in the AI era will be the ones that govern tech adoption with clarity, cohesion, and long-term strategy, not those chasing every shiny new app.

As AI reshapes industries, businesses increasingly adopt AI-powered applications to improve efficiency, enhance customer experience, and drive innovation. However, many organizations fall into a dangerous trap—implementing isolated AI solutions that operate in silos, disconnected from the broader enterprise ecosystem. While these standalone tools may deliver short-term wins, they often fail to drive long-lasting transformation or provide real enterprise value.

If your AI tools are not centralized within your business infrastructure, they are not truly part of your organization's digital evolution—they're simply an expensive tech novelty.

The Fragmentation Problem: Siloed AI Limits Business Growth

Adopting isolated AI solutions often creates fragmented workflows that lead to inefficiencies, duplication of effort, and data inconsistencies. Many businesses rush to adopt AI-driven tools for specific tasks—customer service chatbots, predictive analytics, or document automation—but fail to integrate them into their broader enterprise architecture.

The Consequences of Siloed AI Solutions

- Lack of Data Flow Between Systems: AI applications should not operate as isolated islands. Suppose AI-driven insights and automation do not feed into the rest of your enterprise (CRM, ERP, data warehouses, analytics platforms, etc.). In that case, your business is missing opportunities for deeper insights and holistic decision-making.
- Higher Operational Costs: Running multiple AI solutions separately increases maintenance costs, licensing fees, and complexity, often outweighing the benefits these tools provide.
- Manual Workarounds and Redundant Efforts: Employees often copy and paste data between AI tools and legacy systems, reducing efficiency and increasing the risk of errors.
- Security and Compliance Risks: Siloed AI applications create data governance nightmares—sensitive business data is scattered across multiple platforms with little oversight, weak access controls, and no centralized security policies.
- Vendor Lock-In: Many standalone AI applications rely on proprietary technology that prevents seamless migration to other platforms. If a vendor discontinues a product, businesses are stranded with an outdated tool that cannot evolve.

Are These AI Solutions Built to Last?

We've seen this before. Years ago, many companies invested heavily in custom SharePoint solutions—only to find that many small vendors failed to support their products in the long run. When Microsoft rolled out new versions, these businesses were left with unsupported, outdated solutions that no longer worked. The same is happening now with AI.

Many AI startups are pushing niche solutions that promise automation, machine learning, or advanced analytics, but how many will survive the next five years? If your AI vendor lacks long-term viability, your business risks being left with an unsupported, obsolete tool that no longer integrates with your evolving tech stack.

Flashy AI features do not guarantee long-term success. AI must be scalable, maintainable, and built for interoperability.

Integration is Key: AI Must Be Part of Your Enterprise Ecosystem

Standalone AI applications are not the future—integrated AI platforms are. Businesses that win with AI take a strategic, long-term approach to implementation. This means selecting AI solutions that seamlessly integrate with your organization's existing tech stack and allow data to flow freely between departments.

What True AI Integration Looks Like

- AI that connects with your enterprise systems (ERP, CRM, financial tools, document management platforms).
- AI-driven automation that works across business processes rather than only within isolated tasks.
- Centralized data management, ensuring AI models are trained on clean, enterprise-wide datasets rather than fragmented, department-specific data.

- Scalable AI solutions that can evolve with your company's future needs instead of being locked into outdated systems.
- Interoperability across cloud and on-premises environments allows AI models to process and analyze enterprise data no matter where it resides.

The Future Belongs to AI Ecosystems, Not Isolated AI Products

AI is not just another IT tool—it should be treated as a core part of your business strategy. Businesses must shift their mindset from AI as an application to AI as an infrastructure layer that enhances all business operations.

Avoid AI solutions that don't integrate with your existing ecosystem.
Don't invest in short-lived AI trends with no long-term roadmap.
Don 't be fooled by marketing hype—focus on enterprise AI solutions that offer lasting value.

Invest in AI that will stand the test of time—solutions that adapt, evolve, and integrate seamlessly into your organization's future.

The Cost of Short-Term Thinking in AI

AI is a transformational force, but only if businesses use it strategically. A company running five different AI tools that don't talk to each other is no better off than a company still using manual processes—because, in both cases, the organization is not genuinely AI-driven.

Ask Yourself:

- Does my AI investment integrate seamlessly into my company's broader strategy?
- Will this AI Solution still be viable in five years?
- Am I leveraging AI to transform my business or just collecting flashy tools to say we "use AI"?

If your AI isn't centralized, integrated, and scalable, you're just spinning your wheels. It's time to move beyond fragmented AI experiments and focus on fundamental AI transformation.

Garbage In, Garbage Out: The Hidden Roadblock to AI

What Does GIGO Mean Today?

As organizations have raced to embrace the cloud, many have followed different paths, and not all were well-structured. Some migrated everything at once, often under tight deadlines, planning to clean up their systems later. Others chose a hybrid approach, meticulously handling only business-critical workloads while giving less attention to the rest. A select few already had robust on-premises data practices, making their transition to the cloud relatively seamless.

Most companies fall into the first category, quickly migrating systems to meet a pressing deadline. This "just get it done" mindset results in unorganized data that AI cannot effectively interpret. Workflows may function on the surface but lack the complete, structured data required for accurate AI-driven insights. An ERP system like SAP might carry inaccurate records because data governance wasn't a priority during migration. Over time, systems change, employees leave, and transactional records go missing or remain unused, leading to errors, misinformation, and inefficiencies.

These oversights underscore the importance of proper governance and data hygiene. Poorly governed legacy data introduces inconsistencies that inevitably lead to incorrect calculations, misinformed decisions, and frustrating customers or partners. In the era of AI, such lapses become glaringly obvious—and dangerously costly.

Why GIGO Matters in AI and Automation

Artificial intelligence and automation are only as accurate as the data they receive. If the information in your systems is incomplete or incorrect, expect flawed calculations and misguided decisions.

For example, OCR tools like Microsoft Syntex rely on clearly legible, well-structured data from paper forms to extract meaningful information. AI-driven tools will generate unreliable outputs without high-quality input, making automation efforts inefficient or counterproductive.

Consider event planning: Minor data errors in ticket pricing, vendor details, or sponsor information can quickly escalate into more significant issues—from confused attendees to lost revenue and reputational damage. The key is proactively training AI to understand your desired outcomes while aligning data collection and workflows with business goals, by clearly defining the "why" and "what," you ensure that AI operates effectively and delivers the desired results.

The Importance of a Strong Data Input Culture

Data inputs come from multiple sources—human entries, automated triggers, and integrations—and each source presents unique challenges for ensuring accuracy.

- Do you require specific fields to be filled? Missing data leads to inaccurate outputs.
- Are default values being used appropriately? Improper defaults can create misleading reports.
- Are users trained on why their input matters? Educating employees about the purpose of collecting data fosters more accurate entries.

Governance plays a crucial role in maintaining clean data inputs. AI-driven tools can enforce validation rules, automatically checking for inconsistencies or missing information. Blockchain technology can record specific data entries and validate formats (e.g., ensuring names are spelled correctly or within character limits). Organizations that cultivate a culture prioritizing accurate, purposeful data entry set themselves up for more reliable AI outcomes and smoother, more efficient processes.

Best Practices for Ensuring Quality Input

Using Data Validation Tools

- Implement form validators, automated checks, and other tools to ensure the accuracy and consistency of data before it enters your systems. This minimizes the risk of errors propagating.

Leverage Automation

- Automate repetitive or manual data-entry tasks whenever possible. Reducing human involvement in mundane data tasks reduces typos and inconsistencies, creating a cleaner dataset for AI to process.

Establish AI Governance

- Define clear rules and guidelines for updating, integrating, and monitoring data. A robust governance framework ensures that errors are caught quickly, preventing them from undermining AI initiatives.

Treat Data Quality as an Ongoing Process

- Regularly audit datasets for inaccuracies, missing fields, and outdated information. The cleaner your data, the more effectively AI can deliver accurate insights and recommendations.

Clean Up Legacy Systems

- If old systems contain valuable data, invest time normalizing, migrating, or updating it for modern platforms. Consolidating information in newer databases or data lakes creates a more seamless environment for AI tools.

Centralize Your Data

- The more unified and accessible your data is, the better equipped AI will be to handle complex tasks. Centralized data governance unlocks more efficient, scalable AI applications across the enterprise.

A Thought Shift: Input Is the Foundation for Innovation

Many organizations rush to adopt AI and automation, believing they can't afford to waste time. However, poor data practices—hastily implemented migrations, lack of data validation, and incomplete governance—lead to flawed AI outputs and wasted resources. "Garbage In, Garbage Out" is a preventable problem if data input and quality are treated as essential building blocks rather than afterthoughts.

With accurate data, AI can perform to its fullest potential, delivering informed decisions, more intelligent predictions, and dynamic recommendations. Automation thrives in this environment, reducing manual workloads and increasing efficiency across the company's ecosystem. Rather than struggling with incomplete or outdated information, leaders can access real-time insights on demand, freeing teams to tackle strategic challenges instead of mundane data tasks.

Imagine entering a board meeting and accessing reliable, up-to-the-minute reports with zero preparation. That level of agility and confidence comes from well-governed, high-quality data—because AI is only as good as what you feed it. Recognizing that clean data drives innovation sets the stage for a more competitive, responsive, and successful enterprise.

Call to Action

Now that we understand the importance of avoiding "Garbage In, Garbage Out," it's time to act. Adopting AI successfully means investing in more than technology; it requires robust data practices, a long-term mindset, and a commitment to ongoing governance. Here's how you can take the next step:

- **Assess Your Current Data Quality** – Conduct a full audit of your structured and unstructured data.
- **Consolidate and Streamline Your Systems** – Eliminate redundant platforms and ensure data flows smoothly.
- **Build a Culture of Data Accuracy** – Train employees and enforce data governance best practices.
- **Prepare for Future Growth** – Invest in scalable solutions and AI-ready platforms.
- **Adopt AI with Purpose** – Don't just deploy AI—ensure it's aligned with business goals and quality data.
- **Form a Dedicated Team** – Data governance is an ongoing process requiring cross-functional teams to manage it.

With AI adoption still in its early stages, the window for gaining a competitive edge is wide open. The key is taking a deliberate, thoughtful approach to data management that prioritizes accuracy and consistency. Stop scrambling to fix data errors after they've caused problems. Instead, build a foundation today that will power your organization's AI-driven future.

Choosing the Right Database for AI Deployment
Scalability: Planning for Growth

AI workloads can be unpredictable, with data volumes and processing requirements often growing rapidly. Choosing a database that can scale horizontally and vertically is essential, accommodating increased data loads without compromising performance. Consider cloud-native databases like Snowflake, Amazon Aurora, and Google Bigtable, which offer elastic scalability to match your needs. On-premises solutions like Oracle and Microsoft SQL Server also provide robust scalability options but may require more hands-on management.

Key Considerations:

- Can the database scale handle future data and AI model complexity growth?
- Does the database support distributed architectures for large-scale AI processing?

Why Snowflake Is the Ideal Choice for AI

Snowflake has emerged as a game-changer in cloud data management, offering unparalleled capabilities for managing vast amounts of data across multiple clouds. What sets Snowflake apart is its seamless integration with platforms like Microsoft 365, Power BI, Power Automate, and AWS, enabling cross-platform collaboration and AI readiness.

Snowflake's Key AI Features:

- **Separation of Storage and Compute**: Allows scaling without performance bottlenecks.
- **Native AI Framework Integrations**: Supports AI models like TensorFlow and PyTorch.
- **Multi-Cloud Architecture**: Scales across AWS, Azure, and Google Cloud.
- **Time Travel Feature**: Enables historical data access for audit and recovery.

Usability: Empowering Your Team

A database with an intuitive interface, substantial documentation, and a vibrant community can reduce the learning curve and improve productivity. Snowflake offers SQL-based querying and support for advanced data analysis, making it accessible to both technical and non-technical users.

Key Considerations:

- How easy is it for your team to query and manipulate data?
- Does the database offer a user-friendly interface or support for data visualization tools?
- Is there built-in support for different roles, including non-technical users?

Performance: Ensuring Speed and Efficiency

AI applications often require high-performance databases that handle large-scale data processing with low latency. In-memory databases (Redis), columnar databases (Amazon Redshift), and NoSQL databases (MongoDB) serve different AI workloads. Snowflake, however, provides a balanced approach with fast query execution, structured and semi-structured data support, and intelligent workload optimization.

Key Considerations:

- Does the database meet the performance requirements of AI applications?
- Can it handle real-time analytics and batch workloads efficiently?
- Are there known performance bottlenecks that could impact AI processing?

Centralizing Data: Enhancing Accessibility and Integration

AI needs a unified data source to drive holistic insights. Centralizing your data in Snowflake ensures AI models work with accurate, accessible, and scalable data without redundancy.

Key Considerations:

- Does the database facilitate access across different teams and processes?
- What are the benefits of having a single source of truth for AI models?
- How is data security and governance managed in a centralized environment?

Resolving Data Conflicts: Ensuring Data Integrity

Conflicts arise when multiple processes update the same data. Snowflake's multi-version concurrency control (MVCC) helps resolve conflicting updates, while blockchain-based audit trails ensure data authenticity and accuracy.

Key Considerations:

- How do you ensure that conflicting updates do not compromise data integrity?
- Are there tools or algorithms that can automate conflict resolution?

Environment Management: Testing AI Solutions

Before deploying AI, testing in different environments like development (Dev), testing (Test), user acceptance

299

test (UAT), and production (Prod) are essential. Snowflake simplifies this process with its cloning capabilities, allowing quick deployment of test environments without data duplication.

Key Considerations:

- How do you set up and manage Dev, Test, UAT, and Prod environments?
- What tools are available to manage AI deployment across environments?
- How do you ensure testing environments mirror production conditions?

Tools for Managing Testing and Migration

Automating AI solution deployment is critical. Tools like Jenkins, Docker, Kubernetes, Azure DevOps, and AWS CodePipeline integrate with Snowflake, streamlining migration and testing efforts.

Key Considerations:

- What tools are available for managing AI testing and migration?
- How can automation improve the efficiency of AI deployment?
- Are there integrated platforms for managing the entire life cycle?

Bulk Updates vs. Individual Updates: Optimizing Process Efficiency

When updating AI models and datasets, choosing between bulk updates (efficient but riskier) and individual updates (safer but slower) is essential. Snowflake's multi-cluster architecture enables parallel updates, optimizing efficiency while maintaining accuracy.

Key Considerations:

- Should updates be applied in bulk or individually?
- How do you mitigate the risks associated with bulk updates?
- Are there tools to automate and optimize update processes?

The Future of AI-Driven Data Management

Snowflake is not just a data warehouse; it's the foundation for AI-powered, cross-platform data integration. Whether using Microsoft 365, AWS, or both, Snowflake ensures secure, scalable, and efficient data handling, unlocking AI's full potential.

Recommendations:

- Centralize your data for AI readiness.
- Invest in scalable, multi-cloud solutions like Snowflake.
- Adopt automation and AI-powered analytics for better decision-making.
- Train your teams in AI data management through Snowflake Data Engineering School for long-term success.

By selecting Snowflake as your AI database backend, your organization gains a competitive edge in speed, efficiency, and strategic decision-making. The companies that optimize data today will lead the AI-driven future tomorrow.

Green Computing & Sustainability: Optimizing Storage for an AI-Ready Future

The Problem of Digital Waste: Why Efficiency Matters

Cloud data centers account for 18% of the nation's energy consumption, with projections suggesting this figure could rise to 28% by 2030. While many focus on plastic waste and physical environmental issues, the energy footprint of cloud storage is an even more significant concern.

- Every day, millions of documents are uploaded to SharePoint Online, leading to massive digital waste—redundant files, outdated versions, and unnecessary content that consume vast energy resources.
- The Light Bulb Theory: Every time you upload, synchronize, or store a document unnecessarily, it's like leaving a light bulb in an unoccupied room. Now, it multiplies that by millions of users worldwide; this is why cloud storage demands so much energy.

Organizations must adopt green computing practices to declutter cloud environments and reduce unnecessary energy consumption to combat digital waste.

Optimizing Storage and Document Management in SharePoint and OneDrive

File Types and Storage Strategies

- Not all file types belong in SharePoint. Large files like ZIPs, executables, and video files often inflate storage costs without providing collaborative value.

- Microsoft 365 allows up to 250GB file size limits, but organizations should use Azure Blob Storage for non-collaborative content, reducing storage bloat and energy consumption.

Recommendation: Regularly analyze file storage, flag unnecessary content, and migrate non-essential files to Azure Blob Storage for better efficiency.

The Role of Versioning & Syncing in Cloud Efficiency

The Hidden Cost of AutoSave

Versioning: Hidden Digital Pollution

- By default, SharePoint Online saves 500 versions of a document.
- Autosave creates a new version every 25 seconds, meaning up to 144 document versions may be generated in just one hour.
- Each version is a complete copy, significantly expanding storage use and increasing costs.

One document with 500 versions at 10MB each is 5GB—multiply by 10,000 users – Do the math

Best Practice:

- Limit versioning settings based on business needs.
- Encourage using desktop Office apps to control how often versions are saved.

Syncing & Desktop Performance Impact

- Synchronizing between OneDrive and SharePoint creates continuous updates. Even minor changes (like renaming a folder) trigger multiple sync events across all connected devices.
- Excess syncing strains CPU, disk usage, and network resources, leading to slow performance and unnecessary energy use.

Lesson Learned:

- Synchronize only what is needed. Avoid synchronizing entire directories when only a few files are required.
- It's not just the changes you made that are synchronizing; it's everyone who has access to every document changes you are syncing
- Minimize redundant syncing to reduce user desktop CPU usage & latency issues.
- When traveling, use offline connectivity first to update your changes and then synchronize your changes connecting back to the internet.

Enhancing Collaboration While Reducing Storage Waste

Co-Authoring & Syncing Conflicts

- Real-time co-authoring in SharePoint boosts collaboration but can create versioning conflicts and sync delays.
- Inefficient folder structures and poor metadata use lead to data silos and slow performance.

Best Practice:

- Encourage users to edit documents directly in SharePoint instead of synchronizing unnecessary files to OneDrive.
- Adopt metadata-driven structures for faster search, automation, and AI integration.

Retention, Legal Compliance, and Microsoft Purview

Retention Policies & Digital Cleanup

- Microsoft Purview helps organizations automate compliance & data governance, ensuring that only legally required documents are stored.
- Disposition workflows help delete outdated content, reducing storage costs and compliance risks.

Best Practice: Work closely with legal & compliance teams to define document lifecycle policies and reduce digital clutter.

Azure Blob Storage for Non-Microsoft Files

- For non-Microsoft file types (e.g., ZIPs, executables), Azure Blob Storage provides a cost-effective and compliant solution for storage management.
- This reduces the burden on SharePoint Online, preventing unnecessary storage bloat.
- Note: Microsoft Syntex does not work with non-Microsoft file types except PDFs.

Performance Challenges in Microsoft 365

Latency & Document Processing Issues

- Organizations using split-tunneling VPNs often experience latency in document retrieval & versioning updates.
- Slow synchronization & lost document changes frustrate users and hinder collaboration.

Best Practice:

- Regularly test network connectivity uses tools like Test-Net Connection to identify performance bottlenecks.
- Use monitoring tools to track real-time performance issues & sync delays.

A Sustainable, AI-Ready Future in the Cloud

To prepare for AI-driven automation, organizations must focus on efficiency, sustainability, and governance:

- Reduce digital waste by optimizing storage & removing unnecessary files.
- Use metadata & content types to create an AI-friendly environment.
- Leverage Microsoft Purview & Azure Blob Storage for brilliant data retention & compliance.
- Streamline collaboration by eliminating redundant syncing & excessive versioning.

The Light Bulb Theory:

Just as we wouldn't leave a light bulb on all day without reason, we must manage our digital footprint in the cloud. Every unnecessary file contributes to energy waste & inflated storage costs.

AI Readiness Requires Clean Data

AI cannot function properly in a disorganized digital environment. Organizations must effectively remove digital waste and structure data to unlock AI's full potential.

By embracing green computing, storage optimization, and compliance-driven governance, businesses can achieve AI-driven efficiency while reducing environmental impact.

Best Practices for SharePoint Online and the Evolution of Collaboration

Managing Document Versioning and AutoSave in SharePoint Online

In SharePoint Online, document versioning has become more intricate with the AutoSave feature in the browser-based version of Office. This feature automatically captures a new version every 25 seconds, rapidly accumulating document versions. At this pace, a user can reach the 500-version default limit in just over 3 hours and 28 minutes.

This creates a significant storage challenge, especially for organizations managing large files or documents requiring frequent edits. Unlike traditional systems where changes are saved manually, the automatic nature of versioning means each saved version is a full copy of the document, consuming considerable storage space.

Strategies for Controlling Version Creation and Storage Costs

1. Use Desktop Office Applications for Better Control
   - The desktop versions of Word, Excel, and PowerPoint allow users to manually save changes at their discretion, reducing unnecessary versioning and storage use.
   - Organizations should encourage users to edit documents in desktop applications when working on files that require frequent but incremental changes.
   -
2. Optimize AutoSave and Adjust Version Limits
   - Since SharePoint Online does not allow the disablement of AutoSave in browser-based Office apps, users should leverage desktop apps to control the saving process.
   - Admins can configure versioning settings, increasing limits to 50,000 major and 511 minor versions, when necessary, though this will impact storage costs.
3. Develop and Enforce Storage Policies
   - Implement clear policies guiding how versioning should be managed, including when to archive or delete old versions.
   - Consider setting up retention policies for versioning, automatically purging versions older than a set timeframe unless needed for compliance.

By understanding SharePoint's versioning system and using desktop applications where needed, organizations can reduce storage waste, manage compliance risks, and maintain optimal collaboration efficiency.

Legacy Protocols and the AI Era: Why Modernization is Critical

As businesses transition into AI-driven operations, legacy protocols remain among the most significant vulnerabilities. If your organization has yet to modernize its networking and security protocols, you are already behind. Legacy systems introduce inefficiencies, security risks, and AI-exploitable weaknesses. Without proactive modernization, companies will find themselves increasingly vulnerable to sophisticated AI-powered attacks while missing the enhanced security and efficiency AI can provide.

The Hidden Risks of Legacy Protocols

Many enterprises still rely on outdated network protocols, exposing them to AI-driven security threats. While IPv6 adoption is increasing, IPv4 remains widely used, leaving organizations susceptible to attacks like IP spoofing. Other legacy protocols, such as FTP and outdated SSL/TLS versions, create security gaps that AI-based tools can easily exploit.

Common Legacy Protocol Vulnerabilities

- **IPv4 vs. IPv6:** The exhaustion of IPv4 addresses has led to increased reliance on IPv6, yet many organizations continue to use IPv4, making them vulnerable to address spoofing. AI-powered automation can execute these attacks on a scale.
- **Port 21 (FTP):** FTP lacks encryption, making it susceptible to man-in-the-middle (MITM) attacks. AI-driven sniffers can identify

unencrypted data transmissions and extract sensitive information.

- **Port 443 (SSL/TLS):** Older SSL versions, such as SSL 2.0 and 3.0, contain vulnerabilities like POODLE. AI-based penetration testing tools can exploit outdated encryption methods.
- **Port 80 (HTTP):** Although HTTPS has largely replaced HTTP, many organizations still rely on legacy systems using Port 80, exposing data to interception.

How AI Exploits Legacy Systems

AI-driven attacks evolve faster than ever, and legacy systems are their easiest targets. Key AI-enabled attack methods include:

- **AI-powered IP Spoofing:** AI can automate the creation of false IP addresses to bypass authentication systems.
- **Predictive Vulnerability Exploitation:** AI can analyze network patterns to detect and exploit weaknesses in real time.
- **Data Manipulation:** Unencrypted legacy protocols allow AI-driven bots to intercept and alter data packets, leading to financial fraud and information breaches.

Blockchain's Role in Modern Cybersecurity

As AI-driven attacks become more sophisticated, organizations must rethink how they secure data, transactions, and authentication. Blockchain offers a decentralized security framework that enhances data integrity and prevents tampering.

Using Blockchain to Reinforce Security

- **Decentralized Security Models:** By leveraging blockchain for authentication and data integrity, organizations can eliminate single points of failure.
- **SSL/TLS Authentication:** Blockchain-based Certificate Authorities (CAs) help prevent certificate spoofing, reducing MITM risks.
- **BGP Routing Security:** Blockchain validates Border Gateway Protocol (BGP) routing, minimizing risks of route hijacking.

AI and Blockchain: A Secure Partnership

- **AI for Blockchain Security:** AI-driven analytics can detect fraudulent activities on blockchain networks in real time.
- **Blockchain for AI Integrity:** Organizations can prevent unauthorized tampering with AI decision-making processes by storing AI model validation data on the blockchain.

Case Study: IBM Food Trust

IBM Food Trust combines AI and blockchain to improve food supply chain security. A similar model can be applied to cybersecurity, where blockchain ensures data integrity and AI detects real-time anomalies.

AI's Role in Strengthening Cybersecurity Frameworks

While AI can exploit legacy vulnerabilities, it is also a powerful tool for cybersecurity defense. AI-enhanced security frameworks can detect, isolate, and mitigate threats faster than traditional systems.

AI-Powered Cyber Defense Strategies

- **Real-Time Threat Detection:** AI-driven security tools analyze network traffic and flag abnormal behavior.
- **Behavioral Analytics:** AI tracks user behavior to detect compromised credentials and insider threats.
- **AI-powered phishing Prevention:** NLP-powered AI can identify and block phishing attempts before users see them.

Case Study: Darktrace

Darktrace uses AI to monitor enterprise networks for abnormal activity. In one incident, ransomware infiltrated a company's system, but Darktrace's AI autonomously isolated infected devices, preventing further spread.

AI and Security Statistics

- **90% of security professionals** believe AI is critical to modern cybersecurity (Capgemini Research Institute).
- **Companies using AI-driven security systems** reduce breach detection time by **96%** (Ponemon Institute).

Modernizing Protocols for the AI Era

Organizations must abandon legacy systems and adopt modern security protocols. AI-driven cybersecurity frameworks can mitigate risks associated with outdated infrastructure.

Key Protocol Modernization Steps

- **IPv6 Adoption:** The transition from IPv4 to IPv6 reduces address spoofing risks and strengthens network security.
- **Decommissioning Insecure Ports:** FTP (Port 21) should be replaced with SFTP, and outdated versions of SSL/TLS should be eliminated.
- **Quantum-Resistant Cryptography:** AI and quantum computing advancements necessitate the development of encryption resistant to quantum attacks.

Security by Design: Implementing AI-Driven Protocol Updates

Legacy protocols were not designed with AI-era threats in mind. Organizations must integrate security at the design level, ensuring adaptability and resilience.

Zero Trust and AI-Driven Security

- **Zero Trust Networks:** Legacy authentication models rely on implicit trust. AI-enhanced zero-trust frameworks verify every access request dynamically.
- **AI-powered patch Management:** AI can automate vulnerability assessments and prioritize security updates.
- **Blockchain-Based Audit Trails:** Storing security updates on blockchain provides a transparent and immutable record of protocol modifications.

Case Study: Google Beyond Corp

Google's **Beyond Corp** eliminated reliance on traditional VPNs by implementing a zero-trust architecture where AI continuously monitors user activity.

Emerging Standards and AI-Ready Protocols

Adopting emerging security standards is critical as organizations prepare for an AI-driven future.

Key Emerging Technologies

- **DNS over HTTPS (DoH):** Encrypts DNS queries, reducing the risk of data manipulation.
- **Secure API Gateways:** AI-driven API security tools help prevent unauthorized data access.
- **Decentralized Identity Management:** Blockchain-based identity verification reduces reliance on traditional credentials.

Industry Trends

- 78% of organizations plan to adopt zero-trust security models by 2025 (Gartner).
- Companies implementing decentralized identity solutions report a 40% decrease in credential-based breaches (Forrester).

Key Takeaways: Preparing for the Future

AI, cybersecurity, and blockchain converge rapidly, fundamentally altering traditional security models. Organizations that fail to modernize their protocols risk being left behind.

Actionable Next Steps

1. **Eliminate Legacy Protocols:** Identify and phase out vulnerable systems.
2. **Invest in AI-Driven Security Solutions:** Use AI-powered monitoring tools to detect and mitigate threats.
3. **Leverage Blockchain for Integrity:** Secure sensitive transactions and authentication processes with blockchain.
4. **Prepare for Quantum Computing:** Start transitioning to quantum-resistant encryption.
5. **Adopt a Zero Trust Framework:** Ensure security policies validate every access attempt dynamically.

The shift to AI and automation is inevitable; don't let legacy protocols be your downfall. By modernizing your security infrastructure and embracing AI-enhanced defenses, your organization will be better equipped to thrive in the next era of cybersecurity.

Choosing the Right Database for AI: The Foundation for Secure Automation

As discussed in a previous chapter, AI requires a robust data infrastructure. Choosing the correct database is critical for ensuring AI-driven security automation and high-performance analytics.

Key Considerations for AI Databases

- **Scalability:** AI-driven processes generate vast amounts of data. Choose a database that can manage exponential growth.

- **Real-Time Processing:** AI security applications demand real-time data access and processing power.
- **Integration with AI Models:** Ensure your database natively supports AI/ML workloads.

Popular AI-Ready Databases

- **Snowflake:** Cloud-native data warehousing optimized for AI-driven analytics.
- **Google Big Query:** High-speed data processing with built-in machine learning capabilities.
- **Azure Synapse Analytics:** Microsoft's AI-powered data platform with deep integration into security frameworks.

Modernization Is No Longer Optional

AI is reshaping cybersecurity and data management at an unprecedented pace. To remain competitive, organizations must modernize legacy protocols, adopt AI-enhanced security frameworks, and select the proper data infrastructure. The time to act is now—waiting will only increase vulnerabilities and reduce your ability to leverage AI's full potential.

If you want smart systems, start by cleaning your digital house. AI can't fix your mess—it just magnifies it.

Organizations can create a resilient, future-proof digital environment by proactively updating security protocols, integrating AI and blockchain, and investing in scalable AI-ready databases.

SharePoint gave me the foundation. AI gave me the future. Now, I am ready to design systems that think and respond.

Chapter 9:

What Is True Collaboration, and Why Is It So Hard to Achieve?

True collaboration remains elusive despite the wealth of collaboration tools available today, including SharePoint, Teams, Slack, and cloud-based platforms. Even in the 90s, we were trying to solve collaboration with tools like Lotus Notes. The dream of 'true collaboration' has been sold for decades but the dream keeps outpacing the reality. Organizations often and still struggle to fully leverage these tools due to poor planning, leadership disconnects, and a lack of proper implementation strategies.

Collaboration isn't just about software; it's about people. It's the shared experience of working together toward a common goal, supported by the right tools—not driven by them. Throughout my time at various companies, I've consistently seen teams' default to their preferred communication methods—Slack here, Teams there, and email threads—creating confusion, missed messages, and fractured workflows.

"Shadow Collaboration" - parallel, unsanctioned, or fragmented collaboration channels that form because official tools/processes fail to support the team's needs.

The real issue? A lack of clearly defined and enforced standards. Without an approved list of supported apps, the workplace becomes a digital Wild West, where anything goes, and cohesion suffers. When employees believe they can opt out of shared systems or ignore enterprise standards, collaboration breaks down before it begins.

Here's why we continue to struggle with this battle, and what it's costing us.

The 4Cs Collaboration Health Grid

4Cs	Symptoms of Failure	What to Improve
Clarity	Confused roles, duplicative tools	Document ownership, clarify expectations
Consistency	Tools/processes vary by team	Standardized tool use & templates

4Cs	Symptoms of Failure	What to Improve
Culture	Fear, disengagement, information hoarding	Build trust, recognize collaboration
Communication	Info lost in email/slack, fragmented decisions	Use central platforms, recap decisions

Use this as a diagnostic tool or health check. If collaboration feels broken, look at the 4Cs to identify the weak spot.

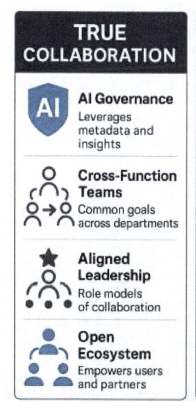
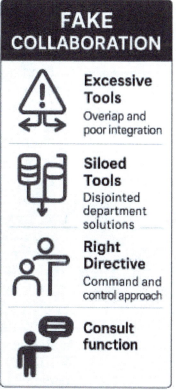

*"Collaboration is more than conversation—
it's about alignment, structure, and purpose."
These visual contrasts real collaboration, built on
governance, shared goals, and open ecosystems, with
the illusion of collaboration—where too many tools,
siloed decisions, and reactive leadership create
confusion instead of clarity.*

Key Issues in Collaboration and Why Many Strategies Fail

1. **Lack of Understanding of Tools and Capabilities**
   - Many employees do not fully understand the tools available, leading to underutilization or ineffective use.
   - Features such as document co-authoring, metadata tagging, and workflow automation go unused simply because users are not trained to maximize them.

2. **Leadership vs. IT in Tool Selection**
   - Business leaders often choose collaboration platforms without consulting IT teams, leading to misalignment between business needs and technical capabilities.
   - The wrong tool choices can create frustration, inefficiency, and resistance to adoption.

3. **Ineffective Project Management and Workflow Systems**
   - Many tools, like Jira and Trello, help with task management but do not actively drive collaboration.
   - Without clear ownership, processes become bottlenecked, requiring excessive

meetings to clarify issues the platform should have handled

4. **Cross-Team Collaboration Barriers**
   - Employees often do not know who is on their team or how to engage other teams effectively.
   - Retrospective meetings often fail to include key individuals, leading to miscommunications, duplicated work, and inefficiencies

5. **Lack of Employee Engagement in Collaboration**
   - Many employees feel excluded, unheard, or undervalued in collaboration.
   - When companies fail to foster inclusive work environments, productivity drops as employees lose motivation and ownership.

Collaboration Maturity Pyramid

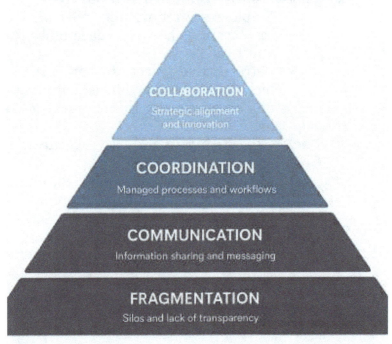

"The Collaboration Maturity Pyramid: From Tools to Transformation"

True collaboration evolves in layers—from basic communication tools to cross-functional alignment, governance, and AI-driven decision support. Most organizations get stuck at the platform level, never realizing that real collaboration is a people and process journey—not just a tech stack.

The Illusion of Collaboration in Corporate Culture

In today's corporate environment, "collaboration" is thrown around like confetti at a parade. It's a buzzword executives love to use, promising innovation, efficiency, and harmony in the workplace. But the reality? True collaboration is rare.

This is more evident than in departmental teams, cross-functional groups, and leadership structures. Too often, individuals bring personal and inter-team conflicts into collaborative initiatives, creating an inefficient and chaotic mess. The core misunderstanding? Collaboration is not about coexistence or the use of software; it's about strategic cooperation.

We layered tech over real values, thinking it would do the work for us. It didn't.

The Disconnect Between Leadership and Reality

One of the most significant barriers to effective collaboration is the disconnect between executives and the employees who execute their vision.

The Know-It-All Effect: How Tech Culture Shapes Growth

The tech industry is filled with brilliant minds—but sometimes brilliance comes with ego. I've worked with incredible mentors who took the time to teach, guide, and challenge me in ways that helped me grow. I've also crossed paths with tech "know-it-alls"—the ones who tear others down to lift themselves up.

Both groups shaped me.

The mentors taught me how to think. The critics? They taught me how to fight—for clarity, for confidence, for relevance.

Some people crumble under that kind of pressure. Some give up. And I understand why—they were not wrong to walk away from toxic cultures or gatekeeping behaviors.

But I chose a different path. I chose to dig in—not out of pride, but purpose.

Because the truth is, every time someone made me feel small, it made me want to build something bigger. Not to prove them wrong, but to prove to myself that I could evolve, adapt, and lead with humility instead of arrogance.

Mentorship matters. So does how we speak to one another in moments of stress or uncertainty. The next great innovator might be watching, listening, and wondering if they belong.

Leadership Starts with Awareness

In 2015, I was fresh off cancer treatment. I had just completed therapy at the Proton Therapy Center in Baltimore and was still on hormone medications known to affect memory. I was commuting long hours every day—MARC train to the Metro—exhausted, still recovering, and doing everything I could to re-enter the tech space with my mind and body still catching up.

I was hired at a high-level agency filled with some of the most brilliant technologists I'd ever worked with. It was the kind of place where expectations were sky-high, and pressure was baked into the culture. That alone would've been a challenge, but I could manage that.

What made it worse was being targeted by someone who should've known better.

He was the classic "tech know-it-all"—but worse. Not just arrogant, but antagonistic. He thrived on undercutting others, making them feel small to feel bigger himself. And for whatever reason, he decided I was his next target.

One day, he walked up to me—no context, no collaboration—and said, "Let me show you something." He pinned the SharePoint Configuration Wizard to the taskbar and set it to run as admin.

"Cool, right?" he said, like he'd just invented electricity.

I nodded and let it go. But inside? I was deeply frustrated. Not because I did not understand what he did but because I did. And I also understood what it meant. He was trying to put me in my place. He did not think I belonged there.

He did not know I was still recovering. He did not know I was fighting to regain my mental sharpness. He didn't know how nerve-racking it already was to work in an elite, demanding environment while not feeling 100%. And honestly, he didn't care.

But I kept showing up.
I kept learning.
I pushed myself to grow. I learned to write PowerShell like code. I contributed to the automation of production deployments. And when I left, I left with my head high—because my work had spoken for itself.

Let's make sure our legacy is more than knowledge, it's how we share it.

How This Shaped What I'm Building

I come from a history of Native American tribes in Virginia and other genealogies. On my father's side, we have deep history in that culture, something I began to understand when I was just a child. I was around six years old when my dad told me that a cousin of ours had broken down on a highway near our home. We went to help, and when we arrived, I saw something I'll never forget: a headrest adorned with feathers. That single image sparked something in me. It made me ask, *"Who am I really?"*

RIP: Aunt Kitchy. A moment. A memory. A mirror of identity.

As I grew older and started exploring more about our people and our history—especially with the rise of the internet—I discovered something profound about Native American leadership. In many tribes, leadership wasn't about privilege. It was about sacrifice. When food was scarce, the leaders didn't eat first—they ate last, after ensuring the tribe was fed. That principle struck me deeply.

This kind of servant leadership—where care comes before command—changed how I looked at business, collaboration, and even myself. It made me ask hard questions: *Do I lead by taking, or by giving? Am I elevating others, or just asserting myself?*

There's a big difference between authority and leadership. The former demands. The latter serves.

These experiences I quote throughout this book didn't just shape how I saw leadership, it influenced what I'm building now.

From every silent struggle, every ego-driven clash, and every moment I felt dismissed, I began to sketch a new kind of system. One that doesn't just track tasks or automate workflows but understands the human side of technology. A platform that values process, governance, and clarity but also respect, inclusion, and awareness.

Tourque is still in development. But what I can say is this:

I'm serving and building something in the tech community that reflects everything I've learned about how people work, what breaks teams apart, and what truly holds them together. Because collaboration without empathy isn't collaboration, it's performance.

And I've seen too much of that.

Leadership Lesson

As leaders, we must create cultures where people can grow—even while they're healing, learning, or quietly struggling. Not everyone comes into the room at 100%. But that doesn't mean they don't belong.

We are responsible not just for what gets done, but for how people feel while doing it. Every great mind on your team is also a human—with challenges, stories, and silent fights. We can't afford to lead with assumption or ego.

While working on a contract for a renowned engineering company, I was brought in to assist with a SharePoint migration project that had been stuck for over three months. The goal was to move more than 15 terabytes of content from SharePoint 2013 to a new SharePoint 2013 environment hosted in Azure. On paper, it was a straightforward migration—from one cloud setup to another—but the process had become anything but simple.

Another technician had already built out the servers and SQL backend, and the team was using Sharegate as the migration tool. Unfortunately, none of the migrations were completed successfully. We were encountering throttling issues, a common problem at the time due to the surge in enterprise cloud migrations— and no full site collections were coming over clean. Days went by with no progress, and despite following the best practices, we were spinning in circles.

The technician had ideas, but nothing was working. When I suggested alternative approaches, I was shut down. Still, I couldn't let the project fail—not just for my own reputation, but because the contracting firm I represented would have been held responsible. So, I decided: I spoke privately with the manager and shared my concerns about the current direction. I asked for the opportunity to try a different approach.

Thankfully, he agreed.

My plan involved bypassing the traditional migration method altogether. I proposed backing up the oversized content databases—which far exceeded Microsoft's best practice limits—and transferring them using AWS S3 tools known for handling large file sets efficiently. One of the content databases alone was over 200 GB, and I was able to move it in just a few hours.

After restoring and attaching the content databases to the new multi-tenant SharePoint farm, I brought one of the largest site collections back online. The site was fully functional.

The next day, I presented the working environment to the project manager. Shortly after, the original technician was released from the project. It was a difficult outcome. I didn't want to see anyone lose their job. But my responsibility was to deliver results, and I had to step up to protect the project and the client's trust.

I ended up finishing the migration almost entirely on my own, with the help of an outside consultant who rebuilt the SharePoint farm based on some key configuration corrections. While he handled that piece, I developed the migration plan, engaged users and managers, organized testing, and executed the rest of the data transfers on schedule.

Reflection & Moral

This experience taught me something I'll never forget: When do you speak up? When do you risk discomfort doing the right thing?

Too often in organizations, people with great ideas stay silent. They fear backlash, office politics, or being seen as disruptive. But when ideas are buried under layers of management or dismissed because they come from someone "lower in the chain," creativity dies.

It's in situations like this that shadow IT is born—because people are trying to solve real problems but are left without support or permission.

Tourque was built to change that. It exists to surface ideas, cut through hierarchy, and empower contributors at every level to move innovation forward. It's designed to prevent organizational gridlock and restore clarity to decision-making and collaboration.

Because sometimes, the person with the solution isn't at the top of the org chart—they're just waiting to be heard.

True leadership is aware. And it protects the potential in people, even before it fully reveals itself.

Leadership Misalignment

Executives often develop grand strategies around governance, best practices, no-code development, licensing, and integration. However, these ideas frequently lack an understanding of the real-world effort required to execute them.

Trust the Experts You Hire

As a consultant, my role has never been just to fix systems to help clients understand *why* things are broken, and what needs to change to move forward. That's the hard part. Change is uncomfortable. And for many, comfort lives in the past.

Over the years, I've worked with many teams who shut down my guidance the moment it pushed beyond what they were used to. One client stands out: I recommend we follow best practices for an upcoming SharePoint migration, including the use of SharePoint groups for clean, scalable permissions. Instead, I was told to "just use the system groups" because it was "easier" and "faster."

It wasn't.

By the time I left, users had access to content they should never have seen. The cleanup was massive and entirely avoidable. If we had done the hard work up front, the management would have been easy. That's what so many teams don't understand: you plan first, so the rest can flow.

If you hire a subject matter expert, let them be the expert. Too many projects fall apart because leaders can't let go of control or insist on suggesting shortcuts based on incomplete knowledge. I've seen decisions made not from strategy, but from distrust—often because someone "had a bad experience with SharePoint" years ago and never moved past it.

And let's be honest: not every manager liked Microsoft products. You could tell by the way they ignored SharePoint's strengths or forced workarounds that didn't make sense. I once had a manager try to port all SharePoint apps into ServiceNow. But ServiceNow isn't SharePoint, and it never will be. SharePoint, with the right implementation, can replicate ServiceNow's workflows. The reverse isn't true.

At the end of the day, this isn't about ego. It's about trust.
Listen to your SMEs. Let them guide you not because they want control, but because they've seen what happens when you ignore the map.

Collaboration Maturity Pyramid

"The Collaboration Maturity Pyramid: From Tools to Transformation"

True collaboration evolves in layers—from basic communication tools to cross-functional alignment, governance, and AI-driven decision support. Most organizations get stuck at the platform level, never realizing that real collaboration is a people and process journey—not just a tech stack.

The Cost of Unproductive Meetings

Throughout my career, I've spent countless hours in meetings that could have been resolved with a brief hallway conversation or a well-designed SharePoint form. The issue isn't meetings themselves but their misuse. Effective collaboration doesn't always require a room; it requires structure, clarity, and purpose.

For instance, discussing detailed requirements to ensure team alignment warrants a meeting. However, addressing a simple task like transferring user rights often doesn't.

I've been in meetings intended to be quick but veered off-topic due to unrelated questions. This diversion disrupts developers, pulling them away from their focus on creating solutions.

In many cases, I've been asked to attend meetings with less technical teams to discuss things like the placement of web parts, site page layout, or even which colors should be used. While design discussions have their place, they often fall outside the scope of what a technical professional is best equipped to contribute. Ultimately, it's up to the customer to decide how they want their pages to look and how they want to communicate—both visually and through their messaging.

I have no control over branding, color choices, or content strategy. I don't know the story you're trying to tell on the page, nor do I direct the tone or purpose of the message you want to portray. Because of that, my ability to help is limited in those areas. Where I can add value is in recommending which web parts to use based on functionality and performance but that's about as far as my expertise extends in design-focused conversations.

The challenge comes when I'm pulled into these types of discussions while I have active development work or data modeling tasks in progress. It becomes a real distraction from the work where I could be making a greater impact. During these times, I had clear technical requirements to complete but was still obligated to attend non-technical meetings. While discussing data, metadata, and columns is essential to solution design, excessive or misaligned meetings can slow progress and reduce overall efficiency.

This issue extends to Agile practices, where I've lost days of focus due to retrospectives and company-wide meetings, sometimes feeling excluded as a contractor.

I'm not against meetings, but resource management is a significant challenge in many environments. Developers should receive requirements and be allowed to focus on their tasks.

The waste of time and money on resources is real. Research indicates that developers at large organizations spend an average of 12.2 hours in meetings each week, compared to 9.7 hours at smaller companies. This "coordination tax" not only consumes valuable time but also fragments focus, making it challenging for developers to engage in deep, uninterrupted work. In fact, engineers at medium and large organizations spend 3.2 more hours per week in meetings than their counterparts at small companies.

Moreover, the constant context switching required to attend these meetings can significantly hinder productivity. Studies have shown that it can take around 23 minutes to refocus after an interruption, such as a meeting. This means that even short, seemingly innocuous meetings can have outsized impacts on a developer's ability to concentrate and produce high-quality work.

To optimize productivity and resource utilization, it's crucial for organizations to critically assess the necessity and structure of meetings. By ensuring that meetings are purposeful, well-structured, and include only essential participants, companies can minimize disruptions and allow developers to focus on their core responsibilities, building and maintaining robust, efficient systems.

Supporting Statistics:

- Employees spend an average of 31 hours per month in meetings, with 50% of that time considered wasted.
- Developers often find meetings disruptive, leading to context switching and reduced productivity.
- It can take around 23 minutes to refocus after an interruption, such as a meeting.

The Reality of Execution: Employees and contractors in the trenches know the technical complexities and challenges leadership often overlooks or underestimates. When high-level decisions are made without technical insight, failure is almost inevitable.

Evolution of Collaboration Tools

This gives a broader lens and emphasizes why the problem persists across eras.

Era	Tools	Challenges
1980s–90s	Lotus Notes, Email	Siloed communication
2000s	SharePoint, Yammer	Poor UI, limited adoption
2010s	Slack, Teams	Tool sprawl, lack of standards
2020s	AI, Automation	No governance, Wild West

Internal Communication Breakdowns

There's also a disconnect between employees, departments, and contractors:

- Employees feel unheard, leading to communication gaps, inefficiency, and resentment.
- Contractors are often treated as outsiders despite their expertise and critical contributions.
- Fear of job security stifles honesty, as employees hesitate to challenge poor decisions.
- This lack of transparency hinders growth, contributes to failed projects, and creates and us vs. them mentality—the opposite of collaboration.

Personality Clashes: The Hidden Barrier to Teamwork

Collaboration isn't just about systems and workflows—it's about people. During my time at Microsoft, I observed how different personalities can either drive or destroy collaboration:

- Type A (Dominant Leaders): Seek control and recognition, often stifling teamwork by making unilateral decisions.
- Type B (Flexible Thinkers): Adapt quickly but sometimes lack assertiveness in driving projects forward.
- Type C (Detail-Oriented Analysts): Focus on precision but risk getting lost in minutiae, slowing progress.
- Type D (Supportive Team Players): They work well in teams but may struggle with decision-making and leadership.

337

In today's AI-driven landscape, collaboration must evolve beyond individual control. It requires a team of problem-solvers, not power-seekers. The best collaborative teams balance these personalities, ensuring ideas are valued over egos.

"Teamwork begins by building trust. And the only way to do that is to overcome our need for invulnerability." Patrick Lencioni

Misusing Social Tools: When Noise Replaces Knowledge

At one of my recent client sites, Teams was being used as the main repository for documenting operational practices. Not SharePoint. Not a wiki. Not even a structured document library. Instead, documented processes—actual workflows and policies—were scattered inside chat threads, buried inside Teams channels.

It made no sense.

The search experience was clunky, unreliable, and completely detached from governance. Crawling through conversation threads to find something valuable wasn't just inefficient—it was a step backwards. You can't manage your business from a chat window.

But this isn't just a one-off problem. It's become a pattern in enterprise environments.

Tools like Viva Engage (formerly Yammer) and Microsoft Teams are increasingly being promoted as the new center of collaboration—intended to replace email, boost culture, and "drive engagement." But in reality, these tools are optimized for interaction frequency, not clarity or value.

We've confused chat with collaboration.

The idea that chat equals productivity has been pushed by the culture-first movement, but that only works when communication is contextual, time-bound, and actionable. Too often, critical decisions, updates, or actions get buried in noisy threads—without any lifecycle management or traceability.

What we end up with is:

- No single source of truth
- No structured way to revisit decisions
- No audit trail or metadata
- And a workplace that rewards noise over signal

From our product perspective, Tourque, this is exactly the problem we set out to fix.

Tourque's Structured Approach to Communication

In the Tourque platform, communication doesn't exist just to chat—it exists to move things forward. We ask a simple but essential question:

"What is the purpose of this communication?" Is it to inform, decide, act, escalate, or reflect?

And then we align the right format to the right purpose, all within context-aware process threads (PCIs).

Communication Type	Viva Engage / Social Chat	Tourque Method
Company Culture / Fun	Great fit – birthdays, fun polls	External to operations. Connected but non-intrusive.
Work Discussions	Messy, hard to track	Tied to Process Instances (PCI) with context and lifecycle
Announcements / FYIs	Lost in noise	Workflow-embedded notifications or dashboard posts
Collaboration & Decision-Making	Scattered across chats	Structured Meeting Notes auto linked to Tasks and PCIs
Knowledge Capture	Not searchable or	Tagged Knowledge Logs

Communication Type	Viva Engage / Social Chat	Tourque Method
	process-bound	with auditability and metadata
Escalations / Reviews	Lost in chatter	Routed Escalations tied to Governance workflows and tracked resolution timelines

This isn't about removing tools like Teams—it's about using them for what they're good at (culture, communication), and shifting critical operations into systems that are structured, governed, and scalable.

Because without that structure, collaboration quickly turns into chaos. And chaos doesn't scale.

The "Wild West" of Corporate IT

Many organizations still operate in a "Wild West" environment, where users do as they please with little governance or oversight. This has long been known as the norm in most organizations unknowingly.

The No-Code Trend has exacerbated this issue, allowing non-technical users to create unregulated applications and workflows. At one large enterprise I consulted for, a team built three separate no-code approaches doing the same thing—because they didn't know the others existed. Imagine how often this happens.

Test solutions proliferate, but these systems become unmanaged, obsolete, or redundant when their creators leave.

IT teams are pressured to implement poorly planned strategies dictated by leadership without questioning feasibility.

When these projects inevitably fail, IT gets blamed despite their warnings. Many professionals fear speaking up, especially when executives choose the wrong technology or vendors due to personal preferences or outdated knowledge.

This is your foundational hierarchy, it shows what to prioritize, and in what order to build a successful collaboration environment.

Breakdown:

1. **People** – The core. Without psychological safety, trust, and inclusion, no tool will help.
2. **Process** – Clear workflows, ownership, and governance. Repeatable methods enable scalable collaboration.
3. **Platform** – Tools like SharePoint, Teams, Confluence, and automation systems should support—not dictate—the process.
4. **AI** – The accelerator. AI enhances everything above *only* if the foundation is solid.

Think of this as your strategic blueprint. AI without process is chaos. Tools without people are wasted. This pyramid is your "build in this order" model.

Despite investments in Microsoft 365 and other modern platforms, many organizations still lack basic governance structures. This leads to:

- No naming conventions or site lifecycle policies in SharePoint, causing confusion, duplicative sites, and abandoned content.
- Over-permission in Teams, where too many users have administrative rights, increasing the risk of data exposure and loss of control.
- Duplicate Yammer or Teams groups with no clear owner, muddying the waters of where conversations happen and who's responsible for keeping them active, secure, or archived.

Without governance, you don't have a digital workplace, you have digital anarchy.

In this environment, collaboration tools become more of a liability than an asset. IT teams are expected to maintain order but are rarely given the authority or support to enforce it. Governance isn't a barrier to collaboration—it's the scaffolding that makes it possible on a scale.

Microsoft 365 Governance Breakdown

A snapshot of why collaboration fails without structure.

Governance Area	Common Failure	Impact
SharePoint Site Naming	No standardized naming or metadata conventions	Confusing site sprawl, hard to find content

Governance Area	Common Failure	Impact
Site Lifecycle Policies	Sites never archived or reviewed	Dead content, increased risk, clutter
Teams Permissions	Too many owners, inconsistent guest access	Security gaps, accidental data exposure
Duplicate Teams/Yammer	Multiple groups for same purpose, no assigned owner	Fragmented discussion, confusion
Power Platform Oversight	No monitoring of apps/flows created via Power Automate/Power Apps	Shadow IT, unsupported business processes
Tool Selection	Teams defaulting to favorite tools with no enterprise standards	Collaboration silos, lost institutional memory

Bottom Line: Governance isn't red tape—it's the foundation for reliable, secure, and scalable collaboration.

When Leadership Ignores Expertise

New executives often overhaul existing systems based on their past experiences without considering the current company's unique needs. This disrupts operations, wastes resources, and frustrates employees.

Instead of building on proven systems, companies restart from scratch—without a solid technical reason.

The result? Unnecessary expenses, operational setbacks, and lower morale.

The Fiction of Collaboration: Why It Rarely Works

The hard truth?

- Collaboration is often an illusion.
- Managers seek more control than they seek collaboration.
- Employees are left out of key decisions and expected to work with tools they didn't choose.
- The voices of technical experts are ignored, resulting in poor technological choices.

This failure is not just frustrating, it is costly. Poor collaboration leads to:

- Wasted budgets on unnecessary software and infrastructure.
- Failed IT projects due to lack of planning and governance.
- Employees disengage, further hindering digital transformation.

The AI Era: Why Collaboration Must Evolve

- The urgency for true collaboration is more significant than ever in the age of AI and automation.
- AI thrives on structured, efficient workflows—something most organizations still lack.

- If companies don't learn how to collaborate effectively, AI will fail to deliver real business value.
- Collaboration is no longer just about teamwork, it's about survival.
- To succeed in the AI-driven future, organizations must:
- Empower IT teams to lead digital transformation efforts.
- Establish governance over automation and no-code platforms.
- Ensure AI solutions are built on well-structured, collaborative foundations.

The Future of Collaboration: A Call to Action

For AI and automation to succeed, businesses must redefine collaboration.

- Executives must trust IT professionals and stop making uninformed technological decisions.
- Employees must be included in the conversation, ensuring tools and workflows meet real business needs.
- Governance must be prioritized to avoid chaos and duplication of efforts.

Find new frameworks that provide a blueprint for this transformation, proving that collaboration is not only possible but essential for the future.

The Collaboration Gap in Agile and ITIL

While Agile and ITIL have defined frameworks for managing projects and services, they lack a built-in collaboration ecosystem that keeps all stakeholders aligned with real-time project data. These methodologies focus on workflows and governance but often fail to integrate with organizations' broader tools for communication, documentation, and project tracking.

Agile: Fast but Fragmented

Agile is widely used for iterative development, emphasizing flexibility and responsiveness. However, it inherently lacks a structured way to store, manage, and retrieve project-critical information across an enterprise. Agile methodologies rely heavily on meetings, stand-ups, and sprint boards often housed in tools like Jira or Azure DevOps, but these platforms are separate from enterprise collaboration tools like SharePoint, Confluence, Microsoft Planner, or Asana.

This results in information silos, where:

- Developers track their work in Agile boards, but business teams lack visibility into project status.
- Requirements and documentation are scattered across different tools, making historical tracking difficult.
- There's no centralized repository to capture project-wide decisions, discussions, and dependencies.

Agile becomes an isolated process rather than a truly connected enterprise workflow without integration into a comprehensive collaboration framework.

ITIL: Comprehensive but Bureaucratic

ITIL provides a structured approach to IT service management, ensuring compliance, reliability, and operational stability. However, its rigid frameworks often stifle real-time collaboration because:

- ITIL prioritizes structured workflows over interactive, dynamic communication.
- Approval processes and documentation requirements slow down adaptability.
- Collaboration happens through ticketing systems rather than fluid, real-time discussions.

Since ITIL does not inherently integrate with modern collaboration tools, teams often switch between multiple systems to track service requests, documentation, and discussions—reducing efficiency and increasing misalignment.

Why This Matters

Project success hinges on more than just process methodologies. Proper collaboration means ensuring that information is centralized, accessible, and actionable across the organization. When Agile and ITIL operate in silos, they fail to:

- Keep real-time project information in one place.
- Provide seamless communication across departments.
- Capture historical context for long-term project visibility.
- Enable teams to make informed decisions without constantly jumping between systems.

To drive efficiency, organizations must go beyond Agile and ITIL and prioritize collaboration-first platforms that integrate project management, documentation, and real-time communication. This ensures that all stakeholders are aligned, not just those following a specific methodology.

Final Thought: The Shift from Fiction to Reality

Collaboration has long been touted as the key to success in modern enterprises, but it often falls short. The assumption that implementing collaboration tools will create a seamless, high-functioning team is flawed. True collaboration isn't about software but people, processes, and leadership alignment.

One of the biggest challenges is corporate control. Leadership often dictates collaboration structures without understanding how work happens at the ground level. Instead of enabling employees, rigid frameworks like Agile and ITIL further entrench silos.

- **Agile** encourages rapid iteration and development but thrives in smaller, focused teams. Once scaled across an enterprise, Agile often becomes fragmented workflows where developers, business teams, and leadership operate in disconnected environments.
- **ITIL**, with its structured service management approach, ensures reliability but lacks agility. It creates bureaucratic bottlenecks that slow innovation rather than empower teams to move forward efficiently.
- **Collaboration tools** like Microsoft Teams, Slack, and Asana are often introduced as "solutions". Still, the organization remains divided if employees' default to their preferred platforms or refuse to share knowledge openly.

The reality is that different personalities play a role in why collaboration fails:

- **The Lone Wolf:** Prefers working independently and resists structured teamwork.
- **The Gatekeeper:** Hoards information, limiting transparency and trust.
- **Passive Observer:** Engages in discussions but takes no initiative to contribute.
- **The Over-Communicator:** Floods channel unnecessary information, making it hard to find what matters.

These factors—corporate control, rigid methodologies, and personality clashes—contribute to the collaboration myth. Many organizations assume that by mandating Agile, ITIL, or a new software tool, they will automatically foster teamwork. However, these efforts fall flat without a culture prioritizing knowledge-sharing alignment and proper integration.

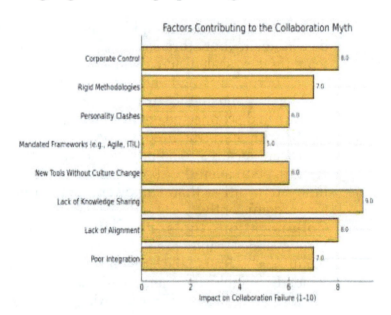

Factors Contributing to the Collaboration Myth

The Path Forward

It's time for corporate leaders to step back and allow the real experts—the people doing the work—to shape how collaboration should function. Agile, ITIL, and software tools should be enablers, not rigid mandates that constrain teams.

Collaboration is not about control—it's about shared success. If organizations fail to embrace genuine, integrated partnerships, their AI initiatives will falter, their IT teams will remain disempowered, and their digital transformation efforts will stall before they even begin.

Understanding Enterprise Infrastructure and AI's Role in Scaling It

As AI-driven processes evolve, organizations must rethink their infrastructure. AI does not work in isolation—it relies on integrated systems, structured workflows, and efficient collaboration tools to be effective.

- Enterprise systems must be optimized for AI adoption, ensuring that workflows, approvals, and processes are AI-ready.
- Companies must identify inefficiencies, such as redundant manual tasks, and streamline them with automation.
- Collaboration tools like SharePoint, Power Automate, and AI-driven analytics platforms can help connect workflows across departments.

The Growing Role of Cloud and Data Centers in AI-Driven Collaboration

The increased automation of business processes means data centers and cloud providers must scale to handle:

- High-speed data transfers for real-time AI processing.
- Expanded storage needs for AI-driven workflow automation.
- Low-latency networking to support seamless collaboration between remote teams.

AI-driven automation has increased reliance on data centers to support intelligent workflow solutions. To ensure AI efficiency, organizations must anticipate network latency issues, workflow failures, and data bottlenecks.

Inclusion as the Key to Collaboration Success

In the AI-powered workplace, inclusion is the most critical factor for success in collaboration.

- Employees must feel valued and placed in roles that align with their strengths.
- AI will be crucial in optimizing workforce efficiency, but human collaboration remains key.
- Organizations must actively restructure their collaboration models, ensuring that:
  - Every team member has a voice.
  - Cross-team engagement is encouraged.
  - Technology is used to complement, not replace, human expertise.

Preparing for the AI-Driven Future of Collaboration

Most companies are not yet prepared for the AI-driven future. Legacy collaboration methods, manual processes, and outdated tools will hold businesses back if they fail to modernize.

- Many organizations still rely on outdated document management practices.
- Manual workflows continue to slow down decision-making processes.
- The lack of AI-driven collaboration tools results in inefficiency and lost opportunities.

To thrive in the future, businesses must:

- Reevaluate their collaboration tools and methodologies.
- Leverage AI-driven workflow automation for faster, more efficient processes.
- Restructure their collaboration models to maximize inclusion and innovation.

Collaboration has always been the steppingstone to automation. The better a company collaborates, the more efficient its automated workflows become. AI will not replace human collaboration but will enhance and streamline turning data into decisions, workflows into intelligence, and ideas into action.

Chapter 10:

AI & Blockchain: The Tech of the Future

In 2020, I was still working as a SharePoint SME and Collaboration Expert, fully immersed in the Microsoft ecosystem. When COVID hit, it didn't disrupt my workflow the way it did for so many others—I had already been working remotely for years. But the world was changing fast, and I felt it. I knew that staying ahead meant evolving with technology. So, I leaned in. I invested my time in learning the Power Platform, diving deep into Power Automate, Power Apps, and Power BI. I even brushed up on my html, CSS, and JavaScript programming skills. That decision paid off. It brought me steady work, new contracts, and a sense that I was on the right path.

But in 2023, everything shifted.

Within six months, I lost every contract I had. One fell apart due to restructuring. Others disappeared, ghosted by budget cuts, economic instability, or a quiet reckoning happening in tech as AI began to steal the spotlight. SharePoint, once the centerpiece of digital collaboration, was being pushed back. I started receiving verbal confirmations for the new contracts—at least ten by phone—only to get follow-up calls days later saying the work was put on hold or canceled altogether. That kind of uncertainty wears on you.

But instead of letting it break me, I let it open my eyes.

I realized I was standing at another turning point. I stood at a crossroads—not of comfort and chaos, but of legacy and leap. One road led backward, patching up old tools. The other beckoned with uncertainty but also with purpose. In mid-2022, I started researching green computing, sustainable storage, and the intersection of AI and business processes. That journey led me to create Tourque, a platform built not only on innovation but also on lessons learned from decades in the field.

More on that to come—but know this: Reinvention isn't just for companies. Sometimes, it's what saves your career. And sometimes, it's where your true impact finally begins.

Events that have shaped our last 5 years:

Year	Key Events
2020	COVID disrupts workplaces, remote work scales
2021	Crypto peak, blockchain exploration
2022	AI buzz begins, NFTs fade, Tourque research begins
Nov-2022	ChatGPT launch – AI revolution
2023	AI chips gain traction, contracts collapse, SharePoint use declines
2024–2025	AI Blockchain adoption, legacy migration, Tourque AI planning

The years from 2021 to the present have been some of the most transformational years in the history of technology. AI is rapidly advancing, the cloud is the foundation of nearly every business, and mobile computing has become more dominant than ever. This era has seen the rise of AI, the evolution of blockchain and cryptocurrency, the collapse of traditional workspaces, and a complete shift in IT priorities due to COVID-19.

The State of Technology (2021- Present)

Three significant trends mark this era:

- The Rise of AI – Artificial intelligence is no longer theoretical; it's here and changing everything.
- Cloud Computing's Total Takeover – Nearly all businesses have moved to the cloud.
- The Aftermath of COVID-19 – Remote work, video conferencing, and automation define the workplace.

Each of these changes how we interact with technology, with people, how businesses function, and even how entire industries operate.

The AI Revolution: How AI Became the Next Big Thing

The Explosion of Generative AI (2022- Present)

While AI has existed for decades, the launch of ChatGPT in 2022 changed everything. This marked the beginning of widespread AI adoption.

- **November 2022 → OpenAI launched ChatGPT**, a generative AI chatbot powered by GPT-3.5.
- **March 2023 → GPT-4 released**, bringing AI to an entirely new level of intelligence.
- **2023- Present → AI adoption skyrocketed**. Every major tech company started integrating AI into their products.

AI no longer requires deep technical expertise to use. Anyone with an internet connection could interact with an AI model, ask it questions, and even generate content. This democratization of AI made it the most significant tech shift since the internet itself.

The Rise of AI Companies

The AI boom led to a surge in companies investing heavily in artificial intelligence. Some of the most prominent players today include:

1. OpenAI → The creators of ChatGPT, leading the AI revolution.
2. Google (DeepMind & Bard AI) → Competing with OpenAI to dominate the AI market.
3. Microsoft → Partnered with OpenAI and embedded AI into Windows, Office 365, and Azure.

4. Meta (Facebook) → Developing AI models for social media, VR, and the metaverse.
5. NVIDIA → The most critical hardware company behind AI, providing GPUs for machine learning.
6. Anthropic (Claude AI) → A rising AI research company competing with OpenAI.
7. Amazon (AWS AI) → Integrating AI into Amazon Web Services to dominate enterprise AI.

AI is still in its infancy, but companies are racing to integrate AI into everything—customer service, automation, business analytics, creative tools, and more.

The Cloud Becomes the Core of IT

By 2021, most companies had already started migrating to cloud, but by 2023, cloud computing had become the backbone of global IT infrastructure.

- **AWS (Amazon Web Services)** → The leader in cloud computing, providing infrastructure for businesses worldwide.
- **Microsoft Azure** → The fastest-growing cloud provider, integrating **AI and enterprise services**.
- **Google Cloud** → Focused on AI-powered cloud services and big data.
- **Oracle Cloud & IBM Cloud** → Competing in specialized enterprise markets.

The Cloud's Impact

- **Businesses no longer need** physical data centers.
- **Cost Efficiency** → Cloud services reduce hardware costs and maintenance.

- **Security & Compliance** $\rightarrow$ More companies moved sensitive data to cloud-based security frameworks.

By 2023, 80% of global businesses were shifted to a cloud-first strategy.

The COVID-19 Impact on IT (2020- Present)

COVID-19 forced businesses to rethink everything about IT, work, and security. Some of the most significant changes included:

Remote Work & The Boom of Video Conferencing

Before COVID-19, most companies had traditional offices. However, after 2020, remote work became the norm.

- Microsoft Teams & zoom experienced explosive growth.
- VPNs and cloud security tools became critical for protecting remote workers.
- Many companies permanently shut down office spaces and moved to a fully remote or hybrid model.

By 2022, remote work was no longer temporary—it became an expectation.

The Shock to Commercial Real Estate

Because of remote work, office buildings became empty, leading to:

- Leasing collapses → Many companies abandoned expensive office spaces.
- Co-working spaces (WeWork, Regus) struggled.
- A shift to flexible workspaces → Companies reduced office footprints, offering employees "hot desks" instead of permanent offices.

This was one of the biggest economic shocks of the decade—commercial landlords struggled, and businesses saved billions by downsizing.

The Economic Disruption

COVID-19 **disrupted global supply chains, fuel prices, and employment**.

- Food & Gas Prices Soared → Fewer employees meant higher costs.
- Massive Layoffs in 2022-2023 → Tech giants cut jobs due to economic uncertainty.
- Automation grew. → Companies invested in AI and automation to replace human labor.

Entertainment - Major Moves

During the COVID shutdown, while independent artists, crews, and small venue owners struggled to stay afloat, major corporations like Live Nation quietly expanded their influence by purchasing distressed venues across the country.

The live events industry, which was brought to a standstill, became fertile ground for consolidation and power grabs. I wrote about this in a crafted an op-ed following Taylor Swift's congressional hearing due to using some of the experiences to create Tourbook my event management application.

Artists' testimonies during the hearing highlighted the dangers of monopolistic practices in the ticketing industry. That meeting wasn't just about a pop concert, it was about control, access, and the future of live events.

COVID-19 revealed just how vulnerable the infrastructure of art and performance truly is. When small venues folded, large players moved in—not to preserve culture, but to dominate it. And unless we rethink how we build systems, whether in events or enterprises, we'll continue handing control to those who already have the most.

The Rise of Blockchain & Cryptocurrency

Cryptocurrency **exploded** in the early 2020s, reaching **all-time highs**.

- **Bitcoin (BTC) hit $69,000 in 2021**, making crypto mainstream.
- **Ethereum (ETH) launched smart contracts**, powering decentralized finance (DeFi).
- **NFTs (Non-Fungible Tokens) exploded in popularity** in 2021-2022.

However, the crypto market crashed in 2022, leading to bankruptcies (FTX, Celsius, and others). Despite the setbacks, blockchain technology continues to grow.

The Current State of Blockchain (2024-Present)

1. Crypto is stabilizing, with Bitcoin & Ethereum leading the market.

2. Blockchain expands beyond finance, and it is used in identity management, supply chain security, and Web3 applications.
3. Governments are regulating crypto more aggressively, ensuring stability.

While crypto saw significant heights and lows, blockchain is still evolving.

The Mobile & App Boom: Mobile is King

Mobile computing **has completely taken over**.

- Smartphones are the primary computing device for most users.
- Apps dominate daily life—from banking to shopping, entertainment, and communication.
- 5G networks have made mobile internet faster than ever.
- Cloud gaming (Xbox Cloud, Nvidia GeForce Now) eliminates the need for consoles.

The shift from PC to mobile is complete—mobile-first experiences drive software development.

The Future: AI & Automation Take Over

Looking ahead, AI and automation will dominate in the next decade. We're already seeing:

1. **AI Replacing Traditional Jobs** → Chatbots, automation, and machine learning replace repetitive work.
2. **The Evolution of AI Assistants** → Siri and Alexa are getting smarter, and AI agents will soon be commonplace in business and personal life.

3. **The Merging of AI & Blockchain** → Secure AI applications using **blockchain verification** to protect against deepfakes and fraud.
4. **Quantum Computing on the Horizon** → Companies like IBM, Google, and Microsoft invest in quantum computers that will change everything.

AI Collaboration Loop

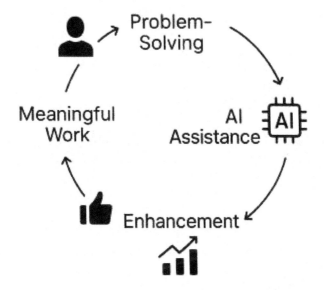

AI Collaboration Loop: *Turning Interaction into Intelligent Action"*
In a mature digital environment, AI enhances—not replaces—collaboration. When paired with structured workflows and governed data, the AI loop becomes a force multiplier: capturing input, generating insights, and driving continuous improvement across the enterprise.

The Reality About AI and Jobs

There's a lot of talk going around that "AI won't take your job"—that it will only "enhance" your productivity or "augment" your role.

Let me be clear: AI *will* take some jobs.
And it won't ask for permission.

We're not talking about sci-fi robots here—we're talking about enterprise-grade automation that can handle workflows, decisions, approvals, communications, and data processing faster than most humans ever could.

With Tourque, if implemented correctly, we predict up to 60% organizational automation and that's not in 10 years. That is *now*.

That doesn't mean the end of human jobs.
But it does mean the end of "business as usual."

Reinvention > Reskilling

People will need to reinvent themselves—not through quick-fix "AI bootcamps" or 8-week programs that promise to make you an "expert."
AI is still in its infancy.
Anyone calling themselves an expert today is more marketer than practitioner.

We don't need more opportunities chasing hype.
We need builders, analysts, architects, and storytellers who can think in systems, data, and human outcomes.
That can't be downloaded in a course. It takes experience.

A Time of Unprecedented Change

Since 2021, the world has seen technological transformation like never before. AI, cloud computing, and mobile apps have reshaped how we live, work, and communicate. Businesses are faster, more automated, and more decentralized than ever before.

This is just the beginning—the next five years will bring even more significant changes as AI advances toward artificial general intelligence (AGI).

The future of computing is no longer about devices, it's about intelligent, connected systems that work seamlessly across all aspects of life.

AI & Blockchain

The intersection of AI, blockchain, and digital transformation offers unparalleled opportunities, but only for businesses that take a measured, strategic approach rather than rushing into adoption. Many organizations struggle with outdated, custom-built in-house platforms that hinder progress, requiring painstaking migrations and restructuring before AI and blockchain can be fully leveraged.

The Hidden Challenges of Migrating Legacy Platforms

Many companies rely on in-house applications—from custom document repositories to niche business tools—that were developed years ago, often by individuals who have since left. Migrating to these systems presents four significant challenges:

- Limited Documentation – Older platforms often lack technical documentation, making migrations guesswork rather than a structured process.
- Dependency Nightmares—Custom platforms frequently have complex, outdated dependencies, which create integration issues with modern AI-driven systems.
- Compatibility Gaps – Many legacy systems were built without scalability in mind, making them incompatible with today's AI and blockchain frameworks.
- Resource Drain and Support – Migration is time-consuming and expensive, requiring expertise in legacy technology and modern AI-driven architectures.

Rather than attempting to recreate outdated applications in a new environment, the best approach is to migrate to scalable, AI-ready platforms that emphasize out-of-the-box functionality. Over-customizing modern solutions is a mistake—it restricts AI adaptability and creates long-term technical debt.

Why AI & Blockchain Need a Unified Data Approach

The biggest mistake companies make in digital transformation is investing in isolated solutions that solve only narrow problems rather than aligning with a long-term strategic vision. These disconnected systems operate in silos, each with its own database, structure, and limited interaction with other platforms. This fragmentation prevents AI from delivering real business value.

For AI to be truly transformative, it needs access to a unified data model that provides a complete business view rather than disconnected data points. Blockchain can play a key role here, ensuring secure, verifiable, and immutable data transactions across an enterprise's entire ecosystem.

AI + Blockchain: The Perfect Combination for Digital Efficiency

- AI enhances blockchain by improving automation, fraud detection, and predictive analytics.
- Blockchain enhances AI by ensuring data integrity, security, and decentralized access, reducing the risks of data manipulation.
- Together, AI & blockchain create a trust-driven digital infrastructure, allowing organizations to eliminate inefficiencies, reduce risks, and automate business processes.

This harmonized approach prevents data silos, ensuring AI models operate on clean, validated, and immutable datasets while leveraging blockchain's security advantages.

A Smarter Approach to AI & Blockchain Adoption

Rather than chasing every new trend, companies should commit to tools that align with their long-term digital vision. At Montego, we emphasize a strategy that:
Centralizes data to break down silos.

- Reducing inefficiencies by integrating AI & automation.
- Enhances security & compliance through blockchain & governance tools.
- Minimizes environmental impact by eliminating digital waste.
- Fosters employee growth with better training & role alignment.
- Sustainability, efficiency, and digital transformation should go hand in hand—not be dictated by the latest fads.

Sustainability: The Quiet Superpower of AI & Blockchain

Beyond automation and analytics, AI and blockchain have another strength: sustainability. When applied with intention, they can:

- Reduce digital waste by automating redundant workflows.
- Lower carbon footprints by optimizing resource allocation and energy use.
- Track sustainability metrics with transparent, tamper-proof blockchain records.

It's not just about digital transformation—it's about responsible transformation. The future isn't only fast, it's cleaner, leaner, and smarter.

AI-Ready Data Management: The Role of Microsoft 365, AWS, and Google Cloud

Microsoft 365: Building a Secure AI-Ready Environment

Microsoft 365 (SharePoint, OneDrive, and Microsoft Syntex) offers powerful AI-driven automation but also presents challenges:

- Document syncing & versioning issues
- Unstructured data impacting AI classification
- Retention & compliance risks without governance

Best Practices for AI Readiness:

- Leverage SharePoint metadata & content types for better AI categorization.
- Use Microsoft Purview for automated data governance & legal compliance.
- Deploy Syntex to convert paper-based documents into AI-readable formats.

Google & AWS: AI-First Strategies

While Microsoft leads in enterprise AI-powered governance, Google Cloud and AWS have their own AI-driven compliance and automation:

- Google Cloud AI & Data Governance – Automated classification & security controls.
- AWS AI & Compliance – Machine learning-driven discovery & retention policies.

By integrating AI and blockchain into governance models, organizations can avoid compliance risks, reduce inefficiencies, and accelerate innovation.

The Path Forward

- AI & blockchain will redefine business models only if data is structured, clean, and secured.
- Companies must transition from legacy silos to unified platforms that support AI-driven decision-making.
- Microsoft 365, AWS, and Google Cloud are all evolving toward AI-first solutions, but success depends on strategy, not just technology.
- Smart adoption of AI & blockchain will separate the leaders from those playing catch-up.

Key Takeaway:

Businesses that rush AI adoption without a solid foundation risk inefficiency, security failures, redevelopment, and poor ROI. The future belongs to companies that invest in structured, AI-ready ecosystems that integrate blockchain, automation, and governance.

AI, Automation, and the End of Legacy Systems

The Case for Portfolio Simplification

One of the most significant challenges organizations face today is the overabundance of tools in their technology stack. Many companies simultaneously use multiple collaboration platforms such as Service Now, Confluence, SharePoint, and Teams.

While each platform may have unique features, their overlap leads to:

- **Data fragmentation** – Information is spread across multiple tools, making it difficult to retrieve and manage efficiently.
- **Inefficiencies** – Employees waste time switching between tools, leading to lost productivity.
- **Security risks** – Unmonitored applications create potential vulnerabilities and compliance issues.

Statistics to Consider

- A BetterCloud report revealed that 80% of employees use unauthorized SaaS applications, creating security risks.
- The average company uses over 110 SaaS applications (Okta Business at Work 2023).
- Employees spend an estimated 12% of their day searching for information across disconnected tools and repositories (IDC).

Strategic Advice for Future Thinking

- Trim Your Portfolio—Identify and consolidate overlapping tools to streamline solutions and make them more adaptable to AI integration.
- AI-Ready Platforms – Invest in platforms with built-in AI capabilities or open APIs for future AI adoption.
- Restrict Shadow IT—Set clear policies to discourage employees from downloading unauthorized applications and use monitoring tools to enforce them.

- Standardized Data Management – Centralize storage and document management to ensure a seamless AI-enabled ecosystem.

At one company, it took me nine months just to obtain a service account needed to run Power Platform solutions. And when I finally received it, it still didn't work, the credentials failed, and the password was set to expire every 15 minutes.

This wasn't due to bad intentions, but rather a deeply complex IT environment plagued by segregation-of-duty rules, legacy identity systems, and an overly siloed portfolio of tools and processes.

What should have been a simple step became an operational blockade. It was a clear example of how process complexity and governance bottlenecks can strangle innovation, even when the right technology is in place.

This consolidation phase is not just about cutting costs; it's about preparing for AI, automation, and next-generation business operations. Organizations that fail to simplify their help desk and tool portfolios will struggle to implement AI-driven processes effectively.

Legacy Systems in the Age of AI: End of the Old Guard

Platforms like SharePoint have been the backbone of collaboration and document management in organizations for years. However, as AI takes center stage, the need for such tools is rapidly evolving. What does this mean for SharePoint and similar platforms?

Why SharePoint and Similar Tools May Become Obsolete

1. **AI-Generated Content** – AI can now create documents on demand, reducing the need for extensive document repositories. AI-powered workflows can automatically draft contracts, invoices, and reports, minimizing reliance on traditional storage.
2. **SharePoint's Storage Inefficiencies** – SharePoint's reliance on versioning and auto-saving makes it a storage hog. Every minor document update creates a full-size version, leading to unnecessary storage costs.
3. **Blockchain for Secure Transactions** – Smart contracts and blockchain-based transactions eliminate the need for static agreements and document-based approvals.
4. **AI as a Corporate Protector** – AI-driven compliance tools can detect intellectual property breaches, prevent unauthorized vendor interactions, and monitor employee behavior for security violations.
5. **The End of Spreadsheets** – AI-generated dashboards can replace Excel, providing real-time analytics without the need for manually created reports.
6. **The Role of Microsoft Syntex** – While Syntex is valid today for converting paper documents to digital content, its long-term value is in preparing data for AI governance.

Preparing for the Shift

Companies must proactively adapt to the AI-driven future by:

- **Adopting Automation** – Transition from manual processes to AI-powered workflows.
- **Consolidating and Simplifying** – Reduce reliance on redundant apps and prioritize centralized AI-ready systems.
- **Preparing for Blockchain and AI** – Explore how blockchain can secure transactions and streamline operations.
- **Embracing Change** – Businesses that fail to adapt cannot compete with AI-first companies.

AI at the Core: How Chips Are Rewriting Computing

The rise of AI chips fundamentally reshapes the computing landscape, and enterprises must adapt their AI applications accordingly. These processors don't just enhance efficiency; they redefine how applications process data, making AI workload faster, more responsive, and significantly more powerful.

AI Chips and the Move to Local Processing

One of the most significant breakthroughs of AI chips is their ability to process data locally, reducing reliance on cloud-based AI models. This shift has several key benefits:

- **Lower Latency** – AI chips execute tasks instantly rather than waiting for data to travel between local devices and cloud servers.
- **Real-Time Decision-Making** – AI-driven applications, such as speech recognition, cybersecurity, and computer vision, respond faster and more accurately.
- **Offline AI Capabilities** – Many AI models can now function without an internet connection, making localized AI solutions a reality.

This shift will impact on how AI models are developed and deployed. Developers should prioritize secure, on-device AI solutions instead of relying solely on cloud-based AI models. These provide faster processing, reduced bandwidth costs, and increased reliability. When combined with blockchain-based security, applications can ensure data integrity while leveraging AI processors for real-time performance.

Performance, Efficiency, and the Power of AI Chips

AI chips provide more than just computational speed; they optimize entire system performances while improving energy efficiency. Key advantages include:

- **AI Workload Offloading** – AI-specific computations are offloaded from traditional CPUs and GPUs, freeing up system resources for other workloads.
- **Lower Power Consumption** – AI processors execute complex calculations with minimal energy, extending battery life in laptops and mobile devices.
- **Optimized AI Workloads** – AI chips accelerate machine learning training and inference, improving overall system efficiency and responsiveness.

For businesses, this translates into cost savings, longer-lasting hardware, and the ability to scale AI-powered applications across multiple devices seamlessly.

Strengthening Security with AI Processors

Security threats are evolving, and AI processors are a powerful defense mechanism against cyberattacks. AI chips provide several key security enhancements:

- **Hardware-Based Threat Detection** – AI-driven anomaly detection identifies malware and security threats in real time.
- **Local AI Processing** – Sensitive data remains on the device, minimizing exposure to external cyber threats.

- **Zero-Trust Security Models** – AI chips facilitate advanced security frameworks that proactively analyze and respond to threats before they materialize.

With outdated security protocols becoming obsolete, AI-powered cybersecurity solutions will detect and mitigate risks faster than ever. Combining AI processing with blockchain security provides a comprehensive defense against emerging cyber threats.

AI-Powered Productivity & Enhanced User Experience

AI chips are set to revolutionize how we work and interact with technology. Here are some of the game-changing improvements:

- **AI-driven Productivity Tools** enhance workflows using instant transcription, translation, and document summarization.
- **AI-Enhanced Media Editing** – Automatic video editing, real-time image enhancement, and AI-assisted design tools make content creation faster and more efficient.
- **Gesture and Facial Recognition**—Touchless interactions and adaptive UI interfaces, like futuristic interfaces in films, provide an intuitive user experience.

Context-aware AI assistants will streamline workflows, making AI an integral part of daily computing rather than just a specialized tool.

Enterprise Applications: AI at the Edge

The healthcare, logistics, and manufacturing industries will benefit significantly from AI processors through edge computing capabilities. AI chips will enable:

- **On-Device Machine Learning** – Eliminating cloud dependencies, enabling real-time AI-powered insights.
- **Smarter IoT Applications** – AI processors enhance industrial IoT, allowing real-time analytics and decision-making at the source.
- **AI-powered robotics** – AI chips will drive autonomous robots in factories, hospitals, and supply chains, improving operational efficiency.

For example, hospitals will deploy AI-powered diagnostic tools for real-time medical imaging analysis, while logistics firms will leverage AI chips for automated inventory tracking and supply chain optimization.

AI in Event Management: Organizing the Future

Event planning is one of the most overlooked yet complex business operations—requiring coordination, logistics, communication, and real-time response. AI is now transforming event management in powerful ways:

- **Smart Scheduling** → AI automates calendar coordination across time zones and departments.
- **Vendor & Budget Optimization** → Predictive tools recommend vendors and forecast event spend.
- **Audience Insights** → Sentiment analysis and engagement tracking in real-time.
- **On-Site Automation** → Facial recognition for check-ins, automated badge printing, and crowd analytics.
- **Post-Event Intelligence** → AI summaries, impact analysis, and follow-up automation.

Whether it's a product launch, internal town hall, or hybrid conference, AI can eliminate stress and replace guesswork with precision. It's not just managing events, it's mastering them.

The Future of AI Creativity & Automation

AI chips are also making their mark in creative industries, transforming how content is generated and edited:

- **AI-Generated Music & Video Production** – AI will automate syncing, editing, and content curation.
- **AI-Powered Graphics & Animation** – Real-time asset generation for gaming and design will reduce production times dramatically.

- **Code Generation & Automation** – AI-assisted coding will accelerate software development and debugging.

While AI streamlines content creation, human creativity remains vital for emotional storytelling and artistic intuition. AI will function as a powerful assistant rather than a replacement for human ingenuity.

AI-powered robotic Process Automation (RPA)

AI processors will enhance Robotic Process Automation (RPA), making business workflows more efficient:

- **Automated Document Verification** – AI will streamline background checks and credential verification.
- **Intelligent Search Capabilities** – AI-powered search functions will enhance document and email retrieval.
- **Organized & Efficient Data Management** – AI chips will optimize data labeling, metadata application, and seamless synchronization across cloud and local environments.

Enterprises must rethink their data structures to support AI-powered RPA effectively, ensuring AI-ready applications and structured data management.

The AI Chip Revolution is Just Beginning

AI processors mark the beginning of a new era in computing. These chips will revolutionize performance, security, and automation across industries by bringing AI processing directly to desktops, laptops, and enterprise systems.

To stay ahead, businesses and developers must rethink how they design applications, manage data, and integrate AI into daily workflows. Companies that embrace AI processors will gain a significant competitive edge, while those that hesitate risk falling behind.

Are you ready for the AI chip revolution? The future of computing is local, intelligent, and lightning-fast—and it's happening now.

The Second Migration: Why Getting AI Right is More Critical Than Ever

We are entering the most pivotal technological shift of our time, the Second Migration.

The First Migration was the move to the cloud, revolutionizing how businesses store and access data. Cloud computing made scalability and accessibility the new normal, but AI demands more. The Second Migration is not just about new tools—fundamentally redefining business operations, decision-making, and efficiency in ways never seen before.

If you do this migration right, your company will evolve with the future. If you do it wrong, you risk becoming obsolete.

Why the Second Migration is Critical

AI can optimize processes, unlock insights, enhance automation, and drive unprecedented innovation. However, without careful planning, AI can amplify inefficiencies, expose security risks, and create technical debt, stifling progress.

Many organizations fail to recognize that AI doesn't just work out of the box. If your data is messy, systems are disjointed, and governance is weak, AI won't magically fix these problems; it will magnify them.

Efficiency, security, and strategic integration are not optional. They are requirements.

From Cloud to Clarity: The First vs. Second Migration

Migration	Focus	Goal	Tech Focus	Risks If Done Poorly
First Migration	Cloud Computing	Scale, flexibility, global access	SaaS, IaaS, virtualization	Vendor lock-in, unmanaged sprawl
Second Migration	AI-Driven Transformation	Intelligence, automation, clarity	AI chips, governance, blockchain	Data chaos, automation fatigue

How to Ensure Your Second Migration Is a Success: Choose the Right Digital Cloud(s)

- AI thrives on data centralization and accessibility. Select platforms that allow seamless integration with AI-driven analytics, automation, and machine learning models.

- Snowflake, AWS, Microsoft Azure, and Google Cloud are top contenders, but the choice must fit your long-term strategy, not just short-term convenience.

Find AI Tools That Fit Your Workforce

- Not every AI solution is right for every company.
- Before investing in AI, evaluate your workforce's needs and select tools that enhance productivity rather than disrupt workflows.
- AI adoption fails when employees don't see the value or face unnecessary complexity.

Secure Your Data from the Start

- AI operates on vast amounts of business-critical data. Without proper governance, security vulnerabilities will skyrocket.
- Implement AI-driven cybersecurity, blockchain for data integrity, and Zero Trust security models to protect against data leaks, unauthorized access, and AI-powered cyber threats.

Develop a Long-Term Data Strategy

The quality of your AI's output depends on the quality of your data. Ensure you have a robust data governance framework that eliminates redundant, outdated, and conflicting information. AI should work with structured, clean, and purpose-driven datasets.

Introducing AI Through Storytelling

AI adoption isn't just about technology; it's about changing mindsets. Executives and employees need to understand AI's role in shaping the future. Storytelling, case studies, and real-world applications can help demystify AI and encourage organizational buy-in.

Offer AI-Relevant Insights on Your Company & Brand

AI is not just a backend tool—it should help shape your company's external messaging, customer interactions, and strategic decisions. Use AI to gain real-time insights into your market, automate customer engagement, and enhance personalization for better user experiences.

Avoid the Pitfalls of AI Adoption

Jumping into AI without proper preparation is a recipe for failure. Many companies rush in only to realize their legacy systems, fragmented data, and disconnected tools make AI implementation a nightmare.

Common Mistakes to Avoid

- Do not declutter existing data and systems before adopting AI.
- Overloading teams with multiple AI tools that don't integrate.
- Ignoring security risks associated with AI-driven automation.
- Failing to align AI investments with long-term business goals.
- Take the time to clean up, optimize, and build a strong foundation for AI. Without this groundwork, AI will become another inefficient, costly experiment rather than a game-changing investment.

The Future Depends on This Move

The cloud prepared us for scale; AI prepares us for intelligence.

AI is not just a tool; it's the next era of computing. The businesses that get this right will lead their industries with faster decision-making, superior automation, and unmatched insights.

- Choose AI solutions that integrate with your long-term strategy.
- Ensuring data governance and security are top priorities.
- Encourage enterprise-wide AI adoption through training and strategic implementation.
- Don't just experiment with AI—make it part of your core business operations.

This is your moment!
The second migration is here. **Are you prepared to leap?**

AI: Autonomy in Action

AI is no longer just an assistant; it has evolved into an autonomous force capable of handling entire business processes. Whether it's automating customer support, managing logistics, or predicting market trends, AI can execute tasks with minimal human intervention. This transition allows teams to focus on higher-level strategy and decision-making rather than getting bogged down in manual workflows.

AI as an Autonomous Operator

Imagine an AI system that monitors and adjusts supply chains based on real-time data, forecasts consumer demand, and makes predictive adjustments— all without requiring constant human input. These possibilities are not futuristic dreams; they are rapidly becoming reality. However, the effectiveness of AI hinges on one fundamental factor: data.

Human-AI Collaboration

AI isn't here to replace human creativity, judgment, or empathy. It's here to enhance them. Human-AI collaboration is where the real power lies—AI handling the repetitive and complex, while humans focus on vision, innovation, and leadership. It's not about replacement. It's about amplification.

Data is the Foundation

For AI to operate effectively, it requires vast amounts of structured, high-quality data. The system must be fed with accurate, curated, and consistently maintained data to learn patterns, make predictions, and drive automation. Without high-quality data, AI remains a sophisticated tool with limited capabilities.

That said, not all data should be freely accessible to AI. Strategic data governance is crucial, as some processes, especially those involving intellectual property, must be restricted. AI should not have unchecked freedom to erase or modify proprietary information, as mismanagement could lead to irreversible errors.

Governance and AI: Protecting What Matters

Governance forms the backbone of successful AI integration. Companies must define where AI can operate freely and where it requires human oversight. Some key governance areas include:

- Data Access – Who controls which data sets AI can use?
- Editing Capabilities – What level of modification is AI allowed to perform?
- Audit Trails – How will AI actions be tracked and logged?

One common concern is data leakage. AI processes vast amounts of information; a poorly managed system could inadvertently expose sensitive data. To prevent this, organizations should implement:

- Encryption – Encrypt data at rest and in transit to minimize security risks.
- Access Control – Limit access to critical data, ensuring only authorized users and systems can interact.
- Regular Audits – Continuously review AI-driven decisions to ensure compliance with data governance policies.
- Data Anonymization – Remove identifiable details where possible to minimize risk exposure.

The Role of Governance in Data Integrity

Strong AI governance policies ensure that AI operates for the business, not against it. Well-managed AI deployments protect intellectual property, maintain data integrity, and prevent unintended consequences. However, governance is more than just a policy; it's a mindset shift. Organizations must foster a culture where teams work together to oversee AI usage, manage exceptions, and continuously refine data inputs.

Creating a Data-Driven Culture

If AI is the engine, data is the fuel. Creating a data-driven culture is essential for AI success. Companies must:

- Educate teams on how AI collects, stores, and uses data.
- Establish clear lines of responsibility for data stewardship.
- Ensure data accuracy and consistency across departments.

Businesses can maximize AI's effectiveness while mitigating risks by embedding a data-first mentality within the organization.

Breaking Free from Outdated Thinking

AI is not just a technological shift but a fundamental change in business operations. Companies must move beyond traditional hierarchies, silos, and legacy decision-making processes.

- Those that adapt quickly, focus on data quality, and build strong governance structures will thrive.
- Those who cling to outdated methods will fall behind.

Transitioning to AI may seem overwhelming for companies with decades of accumulated data. However, success lies in starting small—cleaning up one department or process at a time—while keeping an eye on long-term gains.

AI is not about fitting into a Type A or B personality model. AI adoption requires collaboration at every level of the organization. If companies fail to align their data, processes, and governance strategies with AI, they will struggle to remain competitive in their industries.

The Future of AI in Business

AI is coming, and its influence on businesses will be profound. The key to success is not just adopting AI but adopting AI responsibly. Companies must strategically integrate AI into their business models, ensuring that governance, security, and data integrity remain top priorities.

Organizations that take the time to plan, clean, and structure their data environments today will be the ones who lead the AI revolution tomorrow.

From Survival to Strategy—And into the Future

In May of 2022, I quietly began a different kind of journey—one that didn't start with a contract or a client request. It started with frustration. With curiosity. And with a deep desire to fix what I saw going wrong in the world of technology.

I began digging into ideas like sustainability, efficiency, and collaboration not as corporate jargon, but as principles worth fighting for. I asked myself:

"What would it take to make tech truly better—together?"

The more I searched, the clearer it became that this wasn't just about outdated platforms. It was about decades of disjointed thinking—fragmented systems, siloed teams, and broken governance models dressed up in modern UIs. We didn't need any more tools. We needed to think differently.

Then in 2023, everything I had built professionally began to collapse.

One by one, the contracts disappeared.

At first, it felt like failure. I felt alone buried under bills, with a family depending on me. I didn't know what to do. I tried to fix it myself, chasing new contracts, reaching out to connections, but nothing worked. The frustration grew. The silence got louder. And in that stillness, I realized something deep: I had been trying to do everything on my own. I hadn't trusted God—not really. I was so used to my resume speaking for me, so confident that I'd always land on my feet, that I had forgotten the One who had carried me this far.

At the same time, I had been investing in Tourbook, an event management platform and experimenting with another small app I called Yoke. I believed in the vision, but even those efforts had to stop. I simply didn't have the resources or clarity to keep pushing forward.

Then came the shift.
Clarity.

In the quiet, I prayed and began to hear God more clearly than ever before. He wasn't punishing me; He was preparing me. I wasn't being pushed away from my future. I was positioned for it.

This disruption brought even more focus to the research I had already started: AI, Blockchain, cluttered enterprise environments, and green computing. After helping a large corporation clean up their environment using Microsoft Purview, Syntex, and other governance tools, I doubled down on my investigation in May of 2022.

That journey took me in a whole new direction. Tourbook began to evolve with greater purpose—and eventually, it gave birth to something even bigger: **Tourque**. What started as a quiet season of loss became the catalyst for the boldest platform I've ever built. A product born not out of hype, but out of experience, disruption, and obedience.

It made me step back and re-examine everything—from the systems I'd built to the decisions I'd witnessed in companies going back to the 1980s. And for the first time in my career, I wasn't adapting to someone else's blueprint.

I was ready to build a better one.

*"Be still, and know that I am God." – **Psalm 46:10**

As I researched, I noticed something else. What we see in science fiction often becomes science fact. Take *Total Recall*—that angular electric car? It looks a lot like Tesla's Cybertruck today. Or *Demolition Man*—biometric security, AI assistants, voice-activated homes. We're living in that world now. Movies aren't just entertainment. They're often windows into the mindset of future builders.

Tourque isn't just tech—this is a calling. Tourque was built through brokenness, lessons, and divine insight. I give God the glory for showing me how to turn pain into process, and process into purpose.

The Story Behind Tourque's Creation — Solving What Was Never Truly Solved

For decades, major companies grew by piling layers of software, processes, and technology onto broken foundations.
Instead of solving core problems (governance, structured collaboration, secure information flow), they patched over the chaos with:

- More apps
- More silos
- More manual work
- More "compliance projects" that didn't solve anything long-term

Result?
We built an empire of chaos that looked functional on the outside but was broken underneath.

- Broken collaboration
- Shadow IT
- Poor data structure
- Migration nightmares
- Wasted licensing
- Ineffective cybersecurity
- Talent burnout from bad processes

Tourque was not built to patch the old world.
It was designed because the old world was never built right in the first place.

It's not just disruption for disruption's sake.
It's *correction*.
It's *rebuilding* the foundation.

Tourque stands for:

- Efficiency *by design*, not after the fact.
- Collaboration *by process*, not by chatter.
- Security *by architecture*, not bolt-ons.
- AI *as augmentation*, not a bandaid for brokenness.
- Blockchain *for trust*, not for buzz.

"Unless the Lord builds the house, the builders labor in vain." – Psalm 127:1

Introducing Tourque AI — Born from Experience, Designed for What's Next

Tourque wasn't born in a pitch meeting. It was born in the trenches—in client failures, late-night rebuilds, misaligned leadership, wasted migrations, and collaboration that looked good on paper but failed in practice.

I built **Tourque AI** to solve real-world enterprise problems I've seen repeated for decades:

- Tools purchased before strategy.
- AI bolted on instead of built in.
- Users overwhelmed by automation instead of empowered by it.
- Leadership chasing timelines instead of outcomes.
- Legacy bloat carried forward because no one knew how to let go.

393

Tourque gives organizations permission to leave the clutter   behind —and onboard with clarity.

It's an AI Operating System designed for structured transformation—where process threads, data governance, blockchain-secured actions, and human-machine collaboration aren't ideas... they're built in.

Tourque isn't just another platform, it's a correction to decades of missteps. Here's where Tourque changes the game, and where enterprises will have no choice but to adapt or be left behind.

Where Tourque Disrupts and Redefines

Tourque isn't just another platform.
It's a correction to decades of broken systems, disjointed thinking, and technology for technology's sake.

It's not enough to bolt AI into chaos.
True transformation requires a reset of how businesses think, operate, and collaborate.

Here's where Tourque will change the game:

Disruption Area	Impact
Process Intelligence	Embeds AI into every workflow, not as an add-on but as a foundation.
Blockchain Integration	Brings trust, transparency, and verification to business data exchanges.
Cybersecurity Reimagination	Moves beyond passwords into Zero Trust, hardware

Disruption Area	Impact
	keys, and AI anomaly detection.
Licensing Restructuring	Focuses on modular, scalable licensing tied to *usage* and *outcomes*—not just access.
Governance Evolution	AI-first governance with proactive protection instead of reactive auditing.
Data Management Reinvention	Shifts from "where" data lives to *how* data lives—tagged, validated, and AI-ready.
Collaboration Revolution	Real teamwork structured around outcomes, not noisy, disjointed chat tools.
Sustainability Commitment	Reduces digital waste, optimizes cloud usage, and embraces green computing.

Why This Kind of Disruption Matters

For decades, companies grew on top of chaos, not through the careful solving of problems.

Now, with AI reshaping everything, businesses cannot afford old thinking.

- It's not just about having the newest tools anymore.
- It's about having intentional platforms.
- It's about being data-driven, AI-capable, and collaboration-ready from the inside out.

Embracing meaningful change leads to thriving. Those who cling to the past will be disrupted—whether they are ready or not.

Tourque is not just an innovation. It's restoration. It's the platform built for the next generation of leadership, where efficiency, trust, and creativity come standard.

The future isn't just coming.
The future is already here.
And it's up to us to build it wisely.

Tourbook — Where Rhythm Meets Intelligence

Alongside Tourque is **Tourbook**, a platform inspired by my years in music, performance, and coordination.
Events—like enterprises—require timing, clarity, rhythm, and trust.

From sitting at my dad's baseball games selling snacks, to splicing magnetic tape in a home studio, to managing bands and vendor logistics on the back of napkins, I learned the art of running complex events before I touched a project plan.

Tourbook is not just event management. It's event intelligence.

Built to capture the energy, nuance, and real-time decisions that shape every project, meeting, and milestone.
It brings rhythm to the enterprise—and it reflects a lifetime of creative adaptation.

The Reinvention Loop — And Why It Matters Now

Reinvention is not a one-time thing.
It is a loop. A rhythm. A mindset.
And if I've learned anything—from the field, the stage, and the data center—it's that true leadership lives in that loop.

The willingness to reset. Relearn. Rethink. Reflect. Then rebuild.

That is what separates those who survive from those who lead. Tourque wasn't built from theory. It was built from that loop. And **Tourbook**—well, that's the journal that captures it all along the way.

The Road Ahead

This book is not just a memoir.
It is a blueprint. A systems audit. A human story of grit, grace, and growth through disruption.

If there's one thing, I hope you take from this—it's that the future doesn't belong to those with the most tools.
It belongs to those with the most intentional platforms.

Tourque is my next chapter.
And it might just be yours too.

Because I've seen what technology can do.
I've seen what collaboration can't do when it's unstructured.

And I've seen what AI can become—when it serves people, not the other way around.

With Tourque, and with each other—we can build something better.

Sources and References (for any stats, third-party quotes, or studies)

We can create a "Sources" page that lists:

- BetterCloud SaaS App Report
- Computer World
- Flexera Cloud Report 2023
- Capgemini AI Cybersecurity Study
- Deloitte Digital Transformation Survey
- Golden Steps ABA
- Ponemon Institute Data Breach Report
- Gartner Collaboration Tools Study
- Forrester Reports (as referenced)
- Wikipedia (used as general reference for tech history)
- Business Insider
- Reddit
- A Team
- DevOps.com

Image Credits
ID 51995791 | Mainframe © Everett Collection Inc. | Dreamstime.com

ID 385079 | Mainframe © Bratan | Dreamstime.com

ID 11323180 | Dot Matrix © Viesturs Kalvans | Dreamstime.com

ID 180863587 © Péter Gudella | Dreamstime.com

ID 16619707 © Lapiovradeoctopus | Dreamstime.com

ID 79759656 © Daniil Peshkov | Dreamstime.com

ID 6810357 © Granata1111 | Dreamstime.com

ID 349627553 © Pavlinec | Dreamstime.com

ID 1229608 © Eimantas Buzas | Dreamstime.com

ID 8070 © Photong | Dreamstime.com

ID 225065285 © Roman Belogorodov | Dreamstime.com

ID 253277795 © Bombaert | Dreamstime.com

ID 4831420 © Georgios Kollidas | Dreamstime.com

ID 6008357 © Ivan Korolev | Dreamstime.com

ID 4031815 © Farek | Dreamstime.com

ID 261185024 © Trygve Finkelsen | Dreamstime.com

ID 219139400 © Nicoelnino | Dreamstime.com

ID 4068713 © Matthew Trommer | Dreamstime.com

ID 204220972 © Lacheev | Dreamstime.com

ID 53331078 © Stevanovicigor | Dreamstime.com

About the Autor

Lewin Wanzer is a lifelong technologist, builder, and creator whose career spans over four decades of transformation in the technology industry. Beginning in the mainframe era and evolving alongside some of the most groundbreaking shifts in IT, Lewin has worked across enterprise systems, web development, and AI automation—always with a focus on structure, purpose, and meaningful innovation.

Having a family history of a rich mixed cultural heritage, Lewin draws on heritage, humility, and discipline to guide his leadership approach and his vision for a more intentional digital future. He also brings a unique creative lens to his work, having once been a national-level athlete and professional musician before fully immersing himself in the world of technology.

Lewin has consulted with more than 100 organizations across sectors, helping to modernize legacy environments, train global teams, and architect digital systems that reflect both business logic and human-centered design. His work blends deep technical experience with a passion for organizational clarity and spiritual grounding.

Today, Lewin is the founder of **Tourque AI**, an operating system built to help organizations move beyond siloed tools and toward connected, AI-driven process governance. His story isn't just about systems—it's about people, purpose, and the courage to build what comes next.

My Journey from Mainframe to AI is his first book—part memoir, part guidebook—offering hard-earned insights and practical wisdom for anyone navigating the fast-changing world of technology.

Call to Action

Thanks for joining me on this journey from mainframes to AI.

If this book inspired you, challenged your thinking, or helped you gain clarity on your path forward, I encourage you to stay connected and continue the conversation.

Follow My Work

LinkedIn:
https://www.linkedin.com/in/lewinwanzer/

Website: http://www.tourqueai.com

Learn More

Stay tuned for upcoming blogs, courses, and speaking events focused on:

- AI readiness for organizations
- Green computing and storage efficiency
- Modernizing legacy systems without starting from scratch
- The rise of process integration and AI-led frameworks

Get in Touch

Are you a tech leader, entrepreneur, or consultant interested in AI, cloud migration, or digital transformation?

Let's collaborate or share ideas.

Email: BookInfo@TourqueAI.com